Advances in Range Management in Arid Lands

Advances in Range Management in Arid Lands

Proceedings of the First International Conference on Range Management in the Arabian Gulf

EDITED BY

Riad Halwagy

UNIVERSITY OF KUWAIT

Faisal K. Taha **Samira A. Omar**

KUWAIT INSTITUTE FOR SCIENTIFIC RESEARCH

Kegan Paul International
London and New York
in association with the
Kuwait Foundation for the Advancement of Sciences

First published in 1990 by
Kegan Paul International Ltd
PO Box 256, London WC1B 3SW, England

Distributed by
John Wiley & Sons Ltd
Southern Cross Trading Estate
1 Oldlands Way, Bognor Regis
West Sussex, PO22 9SA, England

Routledge, Chapman & Hall Inc
29 West 35th Street
New York, NY 10001, USA

The Canterbury Press Pty Ltd
Unit 2, 71 Rushdale Street
Scoresby, Victoria 3179, Australia

Set in Times by Kudos Graphics, Horsham, West Sussex
Printed in Great Britain by Whitstable Press, Whitstable, Kent

ISBN 0 7103 0360 2

British Library Cataloguing in Publication Data
International Conference on Range Management in the Arabian Gulf: 1st
Advances in range management in arid lands: proceedings of the First International Conference on Range Management in the Arabian Gulf.
1. Arabia. Gulf states. Rangeland. Development. Management
I. Title. II. Halwagy, Riad III. Taha, Faisal K. IV.
Omar, Samira A. V. Ma'had al — Kuwyt lil — Abh'ath al —
Ilmiyah
333.74'15'09536

ISBN 0–7103–0360–2

US Library of Congress Cataloging in Publication Data
Applied for

The First International Conference on Range Management in the Arabian Gulf

Sponsored and Organized
by
Kuwait Institute for Scientific Research
Kuwait Foundation for the Advancement of Sciences
Environmental Protection Council, Kuwait

ORGANIZING COMMITTEE

Samira A. Omar (KISR)	Chairperson
Wafa' Al-Khamees (EPC)	Convener
Adnan Abdel-Mohsen (KFAS)	Faisal K. Taha (KISR)
Fowzia Al-Sdirawi (KISR)	Sahira Al-Dousari (KISR)

Abdalla T. Omar (KISR)

TECHNICAL COMMITTEE

Faisal K. Taha (KISR)	Chairman
Omar Abu El-Shawarib (AAFRA)	Convener
Samira A. Omar (KISR)	Mohammad A. Razzaque (KISR)
Riad Halwagy (KU)	Adel J. Salman (KISR)

RECOMMENDATIONS AND RESOLUTIONS COMMITTEE

Thadis W. Box	(USA) Chairman
Abdu Al-Sharif	(Saudi Arabia) Co-Chairman
Ahmed El-Naqueeb	(Kuwait)
Fahad Al-Hasawi	(Kuwait)
Faisal K. Taha	(Kuwait)
Mostafa Al-Desouqi	(Kuwait)
Samira A. Omar	(Kuwait)
Salim A. Al-Shanfari	(Oman)
Saad Al-Kuwari	(Qatar)
Abdul-Rahman Al-Dawood	(Saudi Arabia)
Abdel Mawla K. Al-Sulh	(UAE)
J. R. Goodin	(USA)

AAFRA	Agriculture Affairs and Fish Resources Authority, Kuwait
EPC	Environmental Protection Council, Kuwait
KFAS	Kuwait Foundation for the Advancement of Sciences
KISR	Kuwait Institute for Scientific Research
KU	Kuwait University

THE FIRST INTERNTIONAL CONFERENCE ON RANGE MANAGEMENT IN THE ARABIAN GULF

Kuwait, 22–24 April 1985

Contents

FORAGE PRODUCTION AND RANGE ECOSYSTEMS

RANGE ANIMALS

PREFACE

The First International Conference on Range Management in the Arabian Gulf was held in Kuwait on 22–24 April 1985, and was developed with the aim of strengthening and accelerating exchange of knowledge concerning rangelands. The importance of the topic stems from the recent worldwide recognition of the key role that rangelands may play if properly managed and wisely used. This is usually translated in terms of providing goods and services such as forage, livestock, wildlife, water and recreation. Conversely, misuse of the resources has led to devastating results in terms of depletion of vegetation cover, soil erosion and increasing desertification.

The specific objectives of the conference were to: analyze the state of knowledge in range management in the Gulf region and other arid areas, bring into focus the needs for and the role of range research and formulate priorities, highlight the advances made in managing arid rangelands, establish effective communication between people involved in management of Gulf rangeland resources by creating a medium for continuing exchange of view points between them, and provide preliminary recommendations for managing arid rangelands.

To attain these objectives, the Conference Organizing and Technical Committees designed the conference around five technical sessions, one poster session and a field trip to Kuwait's rangelands. Each technical session had a theme and the aim was to treat these themes in depth rather than attempt to address all the important subjects pertaining to management of rangelands. Voluntary contributions to the conference were sought regionally and internationally. Many contributions were subsequently received by the Technical Committee which screened for quality and appropriateness to conference objectives. Twenty-one papers were finally accepted for presentation, of which 17 are published in this volume. Simultaneously, invitations were extended to a number of renowned scientists in range management to make plenary addresses in the five technical sessions.

The conference was held under the patronage of His Highness the Crown Prince and Prime Minister Sheikh Saad Al-Abdullah Al-Salem Al-Sabah. The Minister of State for Cabinet Affairs, Mr Rashed Abdul Aziz Al-Rashed, inaugurated the conference on behalf of His Highness. The conference was jointly sponsored and organized by the Kuwait Institute for Scientific Research (KISR), Kuwait Foundation for the Advancement of Sciences (KFAS) and Environmental Protection Council (EPC). The publication of the proceedings was financed by KFAS.

We would like to express our thanks and appreciation for all the Chairmen of the various sessions of the conference: Mr Nazar Hussain, Dr Mohamed Al-Attar, Dr Thadis Box, Dr A. J. Salman, Mr Fahad Al-Hasawi, Dr J. R. Goodin and Dr M. A. Razzaque. Special thanks and appreciation go to the speakers whose contribution was essential for the success of the conference. We are also grateful to the members of the Organizing Committee, the Technical Committee and the Recommendations and Resolutions Committee who valiantly and efficiently coped with the overall management of the conference and whose unfailing efforts helped so much to make it a success.

Riad Halwagy *Faisal K. Taha* *Samira A. Omar*

RECOMMENDATIONS OF THE FIRST INTERNATIONAL CONFERENCE ON RANGE MANAGEMENT IN THE ARABIAN GULF

Kuwait, 22–24 April 1985

The Participants of the First International Conference on Range Management in the Arabian Gulf, held in Kuwait, 22–24 April 1985, at the Kuwait Hyatt Regency Hotel, wish to express their thanks and appreciation to the Government of Kuwait for hosting this Conference and to the inviting agencies, Kuwait Institute for Scientific Research (KISR), Kuwait Foundation for the Advancement of Sciences (KFAS) and the Environment Protection Council (EPC), for sponsoring the Conference.

GENERAL RECOMMENDATIONS

1. The First International Conference on Range Management in the Arabian Gulf recognizes the significance and importance of rangelands, which constitute more than 90% of the land surface in the Arabian Gulf area. Such resources provide renewable goods and services such as forage, meat, fiber, and recreation. The Conference also acknowledges that rangeland conditions in the region are deteriorating at a rapid pace with such serious consequences as vegetation depletion, soil erosion, and desertification. The Conference therefore recommends that serious efforts be made to realize the true potential productivity of the rangelands. Current efforts in range management in the region should be intensified.
2. The limiting factor in primary productivity in the arid and semi-arid regions is precipitation. Therefore, a total understanding of this precious resource is essential to the design of experiments, the drafting of water management policies, and the efficient use of the water resource. Experiments should be designed with water conservation in mind, and the approach should be one of adapting the plant resource to the existing environment rather than modifying the environment to fit the plant.

SPECIFIC RECOMMENDATIONS

Suggestions and recommendations for developing and improving rangelands in the Arabian Peninsula are listed below.

I. Research Priorities

1. *Plant Research*

   (a) Conduct inventories, surveys and monitoring to show actual and potential range productivity.

(b) Identify desirable species and their economic importance.
(c) Select forage species that are tolerant to salinity and drought.
(d) Screen and develop fodder plants and pastures for feed reserves.
(e) Develop irrigated pasture research.
(f) Evaluate grass response to grazing and stress.
(g) Continue studies on native perennial shrubs and grasses.
(h) Identify, preserve, and develop superior germplasm.
(i) Standardize techniques of measurement and monitoring.
(j) Collect available information and a bibliography on range plants in the Arabian Gulf.

2. *Animal Research*

(a) Conduct studies on grazing behaviour that include palatability of plants, animal movements, and diets.
(b) Conduct nutrition studies on balancing forage supply throughout the year.
(c) Study herd structure and population dynamics of pastoral herds.
(d) Identify and evaluate native and exotic breeds of all animal classes.
(e) Emphasize range nutrition and production of camels, goats, sheep and Arabian horses.
(f) Investigate animal husbandry practices in the rangelands.
(g) Determine population dynamics and the status of wildlife.

3. *Systems Research*

(a) Use integrated grazing systems to assess both plants and animals.
(b) Co-ordinate grazing with ecological factors for maximum sustained yield.
(c) Develop grazing systems that are socially compatible with the traditions of the people and the ecology of rangelands.
(d) Determine grazing capacity and stocking rate.
(e) Determine economic viability of integrated range animal production.

4. *Social/Economic Research*

The arid rangelands of the world support a human population that is either nomadic, semi-nomadic or sedentary. The quality of life for this population must be sustained by stopping further range deterioration, halting or reversing desertification, and addressing the social needs of the people.

The national and regional heritage of people must be preserved. This can be accomplished by a better understanding of the needs of migratory people, the recreational needs of city dwellers, and the aesthetics of the landscape and environmental quality. Therefore we recommend:

(a) A traditional system of grazing use and nomadic movement.
(b) Acceptable techniques for range improvement practices and studies on their economic viability.
(c) Studies on land tenure concepts and status.

(d) Research on such recreational behaviour as:

* Off-route use of vehicles
* Hunting
* Uprooting of plants

II. Recommendations for Development

1. Develop socially acceptable systems for control of grazing (each country).
2. Establish a regional system of grazing control.
3. Set priorities for rangeland uses and co-ordinate activities among such national and regional agencies as municipalities, environmental protection councils, agriculture departments and universities (each country).
4. Develop drought reserves and holding grounds that will allow degraded rangelands to be used seasonally.
5. Rehabilitate degraded rangelands.
6. Utilize appropriate technologies to improve rangelands.

III. Education and Extension

1. Develop a curriculum in range management at some universities in the Arabian Gulf region.
2. Teach range management concepts in technical institutes.
3. Establish demonstration plots on selected sites in all countries to show recommended range improvement practices.
4. Develop short courses on a regional basis, initially at SAKAKA and KISR, in co-ordination with the GCC Secretariat.
5. Send Gulf students overseas for graduate degrees and develop a GCC graduate program in range management.
6. Establish range and animal training centres for Bedouins.
7. Establish public awareness programs at national and regional levels to teach rangeland concepts, and provide Government incentives.

IV. Regional and International Co-operation in Range Management

1. Provide immediate action plan (each country).
2. Hold second Conference in two years.
3. Identify an agency responsible for range work (each country).
4. Develop a directory of institutions, people and research projects in the region.
5. Establish a branch of the Society of Range Management (SRM) or a regional organization for this purpose.
6. Initiate an official mechanism for exchange of genetically improved germplasm.
7. Develop a mechanism for regional exchange of experience and range research information.

Range management needs in the Arabian Gulf region. Keynote address

THADIS W. BOX

College of Natural Resources, Utah State University, Logan, Utah 84321, USA

ABSTRACT

The area around the Arabian Gulf has one of the world's longest histories of man/animal/land interactions. It was near this area that man first domesticated the sheep, goat and dromedary camel. Horses, cattle and buffalo have been known for centuries and are part of the literature and traditions of the people, as well as the biology of the land. In the more recent past, cities have attracted concentrations of animals, and population has grown with the wealth produced by oil. Dust and erosion around the major cities make range management important to urban dwellers and pastoralists alike.

The needs for managing the rangelands are both biological and social. Managers must have a sound estimate of the actual biological production. This is accomplished by measuring what is now there. It is also important to have an estimate of potential productivity — or what can be. When the mechanisms for moving from actual to potential productivity are known, then managers can dare to attempt to move from what is to what can be. This bold move must be acceptable to the people; thus the manager requires a sound sociological assessment of the desires, wants and needs of both the pastoral community and the city population.

Research will play a big role in the development of the Arabian Gulf rangelands. Although inventories of resources and social situations are important, they are not the end product of development. Research will ask questions and test hypotheses on the potential productivity and the mechanisms for reaching that potential. It will be based on good science that relates to principles rather than practices. Its results will be predictable and repeatable.

The range management needs for the Arabian Gulf will be met through sound assessment of the current situations, creative development projects based on existing knowledge, and research based on good science. Ultimately the future of the region may well depend on the quality of science in the research that is conducted.

INTRODUCTION

Range management is the science and art of manipulating native plants and grazing animals to produce long-term sustained yield of goods and services desired by mankind. Although it is generally considered to be a profession of rather recent origin, the Arabian Gulf area is the site of some of the longest man/animal/land interactions in history.

The domestication of the goat and sheep probably took place in the mountains and foothills to the north and east of where we meet today, centuries before any historical records began. The dromedary camel is thought to have been domesticated in the Arabian Peninsula. Cattle, horses and donkeys had all arrived before the first records were kept.

The importance of grazing animals in the life of the people here has been documented in folklore and literature for centuries. Early writings contain extensive passages about herds and flocks, horses and other rangeland animals. Songs and poems of the people in the Arabian Gulf area testify to the importance of livestock and the problems of sustaining them in the harsh environment.

Although we meet today to discuss the modern science of range management, we should be aware that we are doing so near the area where man's husbandry of animals began. It is humbling to think that we discuss research today against a rich traditional background of centuries of living on and managing these desert lands for grazing.

Traditionally, range management has concerned itself with the management needs of the pastoralist or the livestock producer. Much of the effort in this First International Range Management Conference in the Arabian Gulf will address solutions to problems of improving pasturage and livestock. This is as it should be. The pastoralist is the person closest to the land. His livelihood depends on good management.

But another, much larger audience, also has a vital interest in the proper management of rangelands. Large, modern cities now contain the bulk of the people of Gulf countries. These urban dwellers, though their roots may be with pastoral people now live a different life style. Heavy concentrations of animals around cities denude landscapes, increase dust, and in general decrease the quality of life for the urban dweller.

While the goal of improved range management is usually stated in terms of increasing the food and fibre supply for the people of a region, it may be that questions of quality of life may override food needs when the economy provides money for the importation of foodstuffs. For instance, plans for the establishment of the new national park here in the rangelands of Kuwait is an indication of the desire to keep a piece of Kuwait rangeland flora for the people's aesthetic pleasures.

Although range management is primarily an agricultural science that emphasizes animal production, it is also a science that can be used to improve the quality of life for urban dwellers. The conference will concern itself with pastoral needs, but we will be remiss in our duties unless we also address the needs of those who live in the cities.

MANAGEMENT NEEDS

Good land use depends, in part, on three factors: (1) the biological potential of the land, (2) the wants and needs of the people, and (3) the economic and political realities of providing what people want from the land within its biological potential. The needs for proper range management, then, are both biological and social. Most range managers are well trained in the biological sciences, less so in the social areas. However, in both the biological and social arenas the data which managers use may be crude and imprecise, the research information site-specific, land not easily transferred, and projects understaffed. Attempts to manage rangeland on minimal information often leads to their being managed inadequately.

On the surface, the minimal biological needs for managing rangelands appear simple and straightforward. They are: (1) an estimate of actual production, (2) an estimate of potential production, and (3) an understanding of the mechanisms for moving from actual to potential production. In reality, each of these areas is complex. There is a paucity of data and sometimes even ecological theory is lacking, especially in arid regions such as the Arabian Gulf.

Determining actual production is perhaps the easiest of the tasks, but even that task is difficult. The first question is: *what constitutes productivity*? If the answer is obtained by using some plant attribute (e.g. biomass, density or ground-cover), it will be different than if animal products or money is used. Rather than answer the basic productivity question, range managers usually describe the factors relating to productivity regardless of the yardstick used to measure it.

These descriptions are often contained in some sort of resource surveys. These survey reports serve as a baseline for evaluating future management. At a minimum they should include data on: (1) climate, (2) soils, (3) native vegetation, and (4) grazing animal populations.

Climatic data should be intense enough to reflect effective soil moisture content through time. It is especially important to have a measure of variability. Droughts are common in rangelands and the knowledge of the frequency and intensity of these is essential. Rangeland development can best be done during good years, but its ultimate success will be measured by its performance in the drought periods.

Soil should be surveyed in as much detail as possible. Two factors are important to identify: (1) the soil's ability to produce plants, and (2) its stability and resistance to grazing use. The pastoralist is most often interested in increasing feed for his livestock, but the long term productivity is directly related to the soil's ability to resist erosion. The urban dweller may be more concerned with the end effect of wind erosion – dust – than with the animal products from the lands. Development plans should not only consider preventing future erosion, but also healing the scars of past abuse. Good, detailed soil data represent an important tool for the range manager involved in development.

The surveys of natural plant populations should not only describe what vegetation is present, but suggest what plant communities might occur under the best possible management. In order to do this, the survey needs to describe the botanical composition of plant communities along grazing gradients from waterpoints to remote areas, the composition in any relict areas that occur, the intensity of use of each species and the occurrence of rare species. These data must be related to the soils on each range site to get a suggestion of potential plant communities. It should be remembered, however, that at best such exercises only give a suggestion of potential.

The survey of grazing animals will include livestock by species, class, and age. Similar information on native animals and the season when each uses a particular plant community is important. Such factors as nomadic movements of herds and flocks and traditional grazing use areas are especially relevant to the Arabian Gulf rangelands.

Once these surveys are complete, the manager or decision maker has a reasonable estimate of current use and production, but the present conditions are not enough. The suggestion of potential provided by the surveys forms a base for action but the manager needs to know the potential of the land. If improvement is a goal, the mechanisms to move from present to future productivity must be known. In order to provide this information, new knowledge must be developed. Here the research program becomes important.

The potential of an area can never be estimated exactly, but suggestions can be made through several techniques. In newly settled countries such as Australia, Argentina and the United States, range managers have used historical accounts and records to estimate what the land was like before the introduction of livestock. These records, combined with the analysis of relict areas (ungrazed areas such as mesa tops, unwatered areas, etc.) were used to construct an *original* vegetation that was often assumed to be the most desirable. Several serious flaws exist in the technique: (1) the records are inadequate, (2) relict areas are not often typical, and (3) descriptions are usually sketchy at best. The most serious criticism is that the so-called *original* vegetation seldom represents the most productive stage. Even if the technique were without flaws, it would be of little use in the Arabian Gulf area. As noted earlier, a long history of grazing occurred before man recorded his activities in detail.

Another common approach, especially in the computer age, is to calculate potential productivity with mathematical models that consider rainfall, evaporation, soil type, and other environmental factors. This approach has merit, but it is limited both by the lack of descriptive data discussed earlier and by an inadequate understanding of the mechanisms of change in the plant community.

From a practical standpoint, the range manager usually sets an *estimate* of potential based on the best information available. It is important to realize that this is only an estimate and that it should be changed as new knowledge is developed. It is doubtful if the true potential of the Arabian Gulf rangelands will be known in the lifetime of those of us gathered here. It is best to concentrate on the mechanisms of change, to determine how to improve present conditions even if the potential continues to evade us.

Range management has been criticised, correctly, because of its descriptive nature and lack of good science. So-called experiments are often only demonstrations or trials that cannot be properly evaluated. The need to describe current conditions and estimate future potential has led to the gathering of empirical data and much subjective analysis. However, in determining the mechanisms for change, range managers can practise good science. There is an urgent need for the development of theory, the testing of hypotheses and experimentation that can lead to repeatable information. Only then can results be transferred from one site to another. I will return to specific research needs later, but let us look now at the sociological information needed for management of rangelands.

Simply put, the manager needs to know what people expect from the land — their wants, desires and needs. Two major problems arise. The wants must be separated from the needs and needs given priority. Secondly, the audience has to be determined. Here in the Arabian Gulf area there appear to be at least two major audiences: the pastoralists and the urban dwellers.

Pastoralists need adequate nutrition for their animals and themselves. They want to follow their traditional life style, but in addition they want to participate in the amenities of the city. As pastoralists seek the city (their wants), large concentrations of animals occur on the outskirts of most cities, and the nutrition of their animals (their needs) suffers. Lands around cities deteriorate and lose their productivity.

City people certainly need animal products. Most want meat, milk and wool to enhance their life, but they do not want the results of those products being grown near them. If animal products are available for import, they may opt to have a stable vegetation that minimizes dust near their cities and bring animal products from afar — either the more remote areas of their own country or from other countries entirely. City people usually outnumber pastoralists and often have more influence on decision makers in the government.

Thus, rangeland in the gulf region must be managed for both pastoralist and urbanite. Restoration of plant communities to abate dust may be as important as the growing of a forage crop. Development plans should be clear as to what objective they seek. If the goal is dust abatement, unpalatable and grazing-resistant plants may be needed. If providing nutrition for animals during gestation is important, other plants will be used.

Let us now turn to the specific research needs of the Arabian Gulf region. Accurate inventories of physical, biological, social and economic conditions are of utmost importance. They are information-gathering tools for managers and decision makers, and they help form the hypotheses upon which meaningful research develops. The need for more accurate surveys is always present and research for better ways to collect and analyse data is needed. Projects dealing with improved efficiency of surveys should have a high priority. Research into new and better ways of estimating potential is also welcome. However, the most pressing research needs are in the area of understanding the mechanisms of change in rangelands. These include experiments on both the processes of deterioration and the means of improving productivity. At the individual plant level, research needs include studies on the establishment of plants under arid conditions, water stress and drought tolerance of plants, and the resistance of plants to grazing. Study is needed both of the plants native to the region and of promising exotic plants that can be used for forage or for land stabilization.

Animal studies should include grazing behaviour, nutrition, reproduction and performance under rangeland conditions. Individual and flock performance under single species and mixed flocks management is especially important since such knowledge may give the manager a tool for increasing production without degrading the plant resource.

As information on individual plants and animals is developed, another type of research is needed to predict community dynamics and performance. This research on the interaction of plants and animals is the classic area of range research. It includes such areas as the effect of grazing on plant succession, the manipulation of plant communities through animal use, and the establishment of long term sustained yield of rangelands.

Even with the best of management, the rangelands of the Arabian Gulf region will need additional forage production in areas of animal concentration and for drought reserves. Rainfall is low over most of these rangelands. In addition, droughts are common. Research on drought-resistant forage and fodder plants can help relieve the suffering of animals and people during the periods of below-normal rainfall. Improved pastures can provide forage so animals can be removed from depleted rangelands.

Since its inception, range management has advocated: (1) grazing the range with the proper kind or mix of animals, (2) grazing the range in the correct season, (3) distributing the animals well over the range, and (4) grazing with the proper number of animals. These old rules are still valid. The problem is that there is not enough information to allow the manager in the Arabian Gulf to effectively practise them. Information available is mostly empirical, often anecdotal, and usually site-specific. Good science geared to understanding the mechanisms of change in arid areas is badly needed. Experiments that test specific hypotheses to produce information that is predictable, repeatable and that can go beyond a specific site, may help move range management from a descriptive to a predictive science. More work is needed that relates to principles rather than practices.

It is common for an academic to say that more research is needed, but today many Arabian Gulf rangelands are overgrazed, blowing dust settles in major cities, and productivity is low. In practice we may not be able to wait for the good science we know is needed. What can be done now with the development of rangelands?

First, each country can assess its current status. Much of this work is already done; other information can be gathered quickly. The situation will undoubtedly show that there is much more needed than can be done with personnel and funds available. Therefore, priorities have to be set and specific projects selected.

In many cases these projects will involve the conservation of critical sites that may be in the process of degradation, or the rehabilitation of those already degraded. In others they may be directed at improving rangelands in order to balance productivity and use. Whatever the case, sound objectives can be met and action taken on the basis of knowledge available here and overseas. Gaps in the existing knowledge can be filled with research.

Most of the land of the Arabian Gulf region is range. It has been grazed for centuries and it will continue to provide feed for animals. Its ability to produce in the future will depend on the application of the results of good research.

The research conducted by KISR (Kuwait Institute for Scientific Research), and that reported by others at this Conference, represent a step in the right direction. I think you will be favourably impressed at the work you will see in the field tomorrow. It is this kind of research that will protect and improve well over 90% of the lands of this region.

This First International Conference on Range Management in the Arabian Gulf is an important event in the development of the area's rangelands. Speakers here will outline what is known about range management in arid lands during the next few days. Many ideas about how to better manage the Gulf's rangelands will emerge. Specific suggestions for the improvement of the production will be made. It will represent a major synthesis of what is known about managing arid rangelands.

I suggest that the most important accomplishment of the Conference may not be in a synthesis of what is known about the Arabian Gulf rangelands, but a more thorough understanding of what is not known. Much useful development can and should proceed on current knowledge, but the hope for the future depends on the development of new knowledge that allows us to reach the potential of arid areas more easily. It is because of this I say that the future of arid rangelands — and indeed the future of range management as a profession — depends on the quality of science we produce.

Ecology and modeling in relation to rangelands of arid and semiarid regions of Wyoming, USA

HERBERT G. FISSER

Department of Range Management, University of Wyoming, Laramie, WY 82071, USA

ABSTRACT

Monitoring of vegetation, grazing, and environmental factors in semiarid regions of western Wyoming, USA, was initiated in 1960. Plant community dynamics data have been recorded annually at several locations for more than 20 years. On-site seasonal precipitation, temperature and detailed soil profile data have been recorded.

The intent of this report is to portray the ecology of arid land shrub-dominated rangeland vegetation, especially as related to precipitation dynamics, and to identify results of predictive modeling programs.

Attempts to correlate annual herbage yields of arid and semiarid rangelands with total annual precipitation have not often been successful. Use of seasonal precipitation, or combinations of several growing-season month records, have helped improve the apparent relationship between plant growth and climate. This has been useful in Wyoming where growth occurs primarily during spring. Only during exceptional years has summer or fall precipitation modified herbage development. Correlation of biomass production with moisture exhibited good linear relation during most years among sites of similar precipitation, temperature, and vegetation class. In some instances on saline sites, herbage yield was best correlated simply to spring precipitation.

Several linear regression predictive models were investigated. In general if was observed that best precipitation-production relations were noted when only winter and spring moisture values were utilized. The site-specific character of precipitation-herbage yield index procedures often requires adjustments to account for major environmental differences.

Preliminary tests of long-term yield records from the semiarid sites in Wyoming have been conducted with a climate water balance model (ERHYM). Model-predicted soil water and soil temperature were in good agreement with field measured values. With improved model parameters and state values such as crop coefficient and transpiration coefficient, the model could be improved to increase forecasting accuracy and to enhance application of this management-oriented tool to other regions of the world, including the Arabian Peninsula.

INTRODUCTION

If range managers could accurately predict annual forage from their rangelands, they might well be able to avoid total disaster or at least minimize drought impacts by modified management in terms of grazing location and animal numbers. Many regions of the world lack definitive climate-yield information necessary to develop any kind of model. Many livestock managers, however, throughout the developing world and the United States as well, are cognizant of annual climatic changes and often are knowledgeable about the potential for change from one location to another.

The science of range management involves investigation of rangelands and grazing animals. The results of successfully applied range management are stabilized vegetation and soils and, very importantly, improved quality and quantity of meat production for human consumption. The importance of scientific data and knowledge, however, can be realized only if they are applied in the management and use of these lands. The long term research program in Wyoming was developed specifically to generate quantitative and scientific knowledge of rangelands and grazing interaction as well as to provide a data base for application to the often very difficult part of management and use of these lands for grazing animal production.

Most rangelands of Wyoming are dominated by shrubs (Beetle 1960, Porter 1960, Kuchler 1964). The most common and widely dispersed shrub is big sagebrush (*Artemisia tridentata*) (Beetle and Johnson 1982).

Wyoming shrub ranges have supported a viable livestock industry of cattle and sheep for over 100 years (Nelson 1898, Vale 1975, Kearl 1980). Their management and that of very extensive pronghorn antelope (*Antilocapra americana*), mule deer (*Odocoileus hemionus*), and elk (*Cervus elaphus*) populations have had to become more intensive and efficient to meet ever increasing competition for land use and economic stress in private and public uses of rangelands (Kearl and Freeburn 1980, Olson and Gerhart 1982, Heady 1984). Ecological research concerns for sagebrush rangelands were early addressed by Robertson (1947) and Passey and Hugie (1962). More recent efforts by Johnson (1969) and Johnson (1979) exhibited increasing awareness of the continuing importance of sagebrush rangelands to agricultural, wildlife, and energy development industries. This has been exceptionally important in Wyoming with its great vistas of these aromatic plants.

The ecological studies reported here were initiated during the late 1950s (Fisser 1962, 1963). Various aspects of this research have continued to the present. Many ecological and management-based interests have been considered during the years of monitoring on more than 100 locations encompassing the semiarid shrublands of western Wyoming (Fig. 1).

The intent of this report is: (1) to portray the ecology of semiarid shrub-dominated rangeland vegetation, especially as related to precipitation

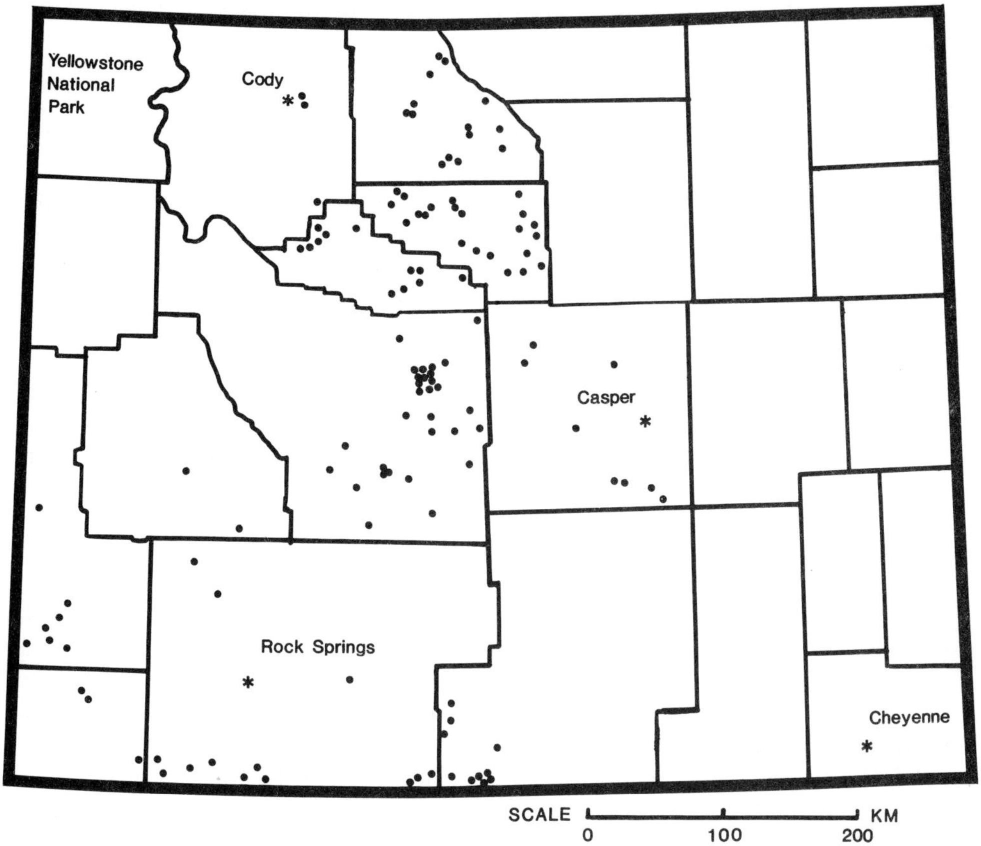

Fig. 1. Map of Wyoming, USA, showing locations of long-term vegetation, soil, and climatic semiarid shrubland research sites.

dynamics; (2) to identify results of predictive modeling programs with simple regression statistics; and (3) to apply a hydrology-based forage production model to investigate validity of prediction and forecasting processes.

REVIEW

Nearly one-half the earth's surface is grazing land of some category including grassland, woodland, shrublands, desert or tundra (Stoddart *et al.* 1975). Rangelands produce a very large proportion of the food consumed by livestock in all parts of the world and livestock meat derived from grazing lands produces protein supplies of extreme importance to many. The steady supply of rangeland forage and the livestock product derived by grazing animals, on an annual basis, is often uncertain. Drouths like those of Africa's Sahel region in 1968–1973, of the Ethiopian area of 1983–1984, and even of the US Great Plains in the 1930's are endemic and to be expected.

Drouths occur in all semiarid and arid rangelands at more or less regular intervals (Cook and Sims 1979, Owen 1979).

A plant community or range site is a complex of vegetation species dependent upon variations in the climax vegetation potential, the climate, the soils, topography and other environmental factors. Soils often assume an almost equal role with that of climate and vegetation in determining the limits of a plant community. The general statement, however, that climate can serve as an indicator of either soils or vegetations often may be incorrect. Plant species may, and often do, vary due to migration of flora and differing parent materials of soil formation. Such factors lead to floristic and structural differences in vegetation, even within the same regional climatic zone (Sneva and Britton 1983).

Climate is usually the most important factor that determines vegetation type as well as annual herbage yield. The local climate, which is largely controlled by the regional climate, may differ greatly from another local climate only a short distance away. The climate of a small area can be affected by changes in elevation or even a local shaded area. Consequently, a description of the climate of a region seldom indicates the environment of a particular plant or group of plants (Moss 1964).

Fluctuation in precipitation is the most important factor causing annual variation in range forage production. The growth of range plants is markedly influenced by both annual and seasonal differences in precipitation and temperature (Whitman and Haugse 1972). Fortunately, indigenous range plants are often highly drought resistant and have become adapted to a particular xerophytic climate due to the process of natural selection. Levitt (1964) indicated that drought tolerance of plant species appeared to be the most primitive characteristic, while avoidance, the most advanced. Xerophytic plants must possess the following adaptations: the ability to survive, grow, flower and fruit, and reproduce under drought conditions. Plants with the highest drought resistance may not produce the most herbage. Most successful "water savers" usually sacrifice yield since closure of stomata leads not only to the saving of water, but also to the cessation of photosynthesis. Only when drought is extreme enough to cause injury to the "water spenders" but not "water savers" can the latter be expected to outproduce the former (Levitt 1964). Daubenmire (1956) indicates that the quantity of water available to plants is not indicated by precipitation alone. The effectiveness of a given amount of precipitation is largely influenced by temperature and other conditions which affect the rate of its return to the atmosphere by evaporation and transpiration.

In addition to the very complex aspects of plant yield interactions with environmental and climatic factors, is that related to plant phenology, the expression and identification of growth and development of plants. Phenological records have been kept for agricultural crops for many years. Only recently have scientists begun to investigate the phenology of native species. The range scientists must not only understand the biotic and abiotic factors favourable for plant growth, but must also interpret properly the impact of those unfavourable. The problems of studying plant growth and associated environmental factors thus become enormous. The enormity of the task is well stated by Goodall (1970):

"To study the end effects of environmental factors on processes or parameters of the ecosystem as a whole is a formidable task. Since the environment is involved in, and a part of, the ecosystem, the problem cannot always be given any clear-cut interpretation. Cause and effect are mingled, often almost inextricably, and while the concurrent changes in biotic and abiotic factors can be described, it may hardly be possible to interpret the network of relations in terms of chains of causation."

Plant phenology is a definitive expression of climatic and environmental interaction of biological elements. It is important when interrelated with environmental phenomena, vegetation production and management of rangelands. Because of the interrelationships of time and rate of plant development and the necessity for resource management decision making in terms of interpretive evaluation and application of prediction and forecasting processes, it becomes extremely important for researchers to clearly understand the general relationships of herbage yield, phenology and weather variations (Kinucan and Fisser 1984).

Phenology has been described by Lieth (1970) as a science integrating floristiscs, ecology, meteorology and observation of life cycles and phenometry as the quantitative analysis and correlation of the same components. To divide the physical environment into discrete components, it is necessary to identify probable plant respose characteristics. Environmental factors often interact and their simple effects also cannot necessarily be isolated. Separate treatment of individual factors has often, however, been convenient and worthwhile because it allows the researcher to develop some overview concepts of the general biotic-abiotic relationships existing on the area of his research.

Quantitative development of models which might represent a potential tool to be utilized in forecasting impacts of future climatic changes are of significant value. The concept of models may relate to extremely complex mathematical systems or can be interrelated with minimal climatic and environmental input. Virtually hundreds of mathematical models, many involving only simple equations, have been developed (Hinds and Van Dyne 1980). The general relations of vegetation response to differences of environmental and climatic changes through time have been considered in great detail. Few, if any, models have been able to adequately estimate forage production for long periods of time and little has been done to determine how a model might be developed and characterized such that its utilization could be extrapolated to different areas of the world. A few models can estimate forage production for reasonably large regions of rangeland. Differences in performance of practical models claiming regional or global generality and requiring only readily available data such as precipitation and temperature may not be able to precisely estimate potential forage yields, but they have the inherently important factor of being significantly useful because of the minimum requirement for data generalization.

Predictions with statistical regression models have been utilized for a number of years and are considered reasonably powerful tools for estimating variables such as forage production. They have the ability to take into account factors such as vegetation, climate and soils for determination of the factor or factors which most closely approximate the potential causal relationship of vegetation, yield, and climatic interactions. They do have a disadvantage when applied to biological systems, however, in that their coefficients remain constant while the factors interact continuously in nature (Kinucan and Fisser 1984).

DESCRIPTION OF RESEARCH SITES

Location and geology

Long-term herbage production has been monitored on more than 40 sites since the early 1960s. Data from two locations have been selected for presentation in this report. The name, county location, legal description, and year of establishment of the two selected research sites are given in Table 1. Smilo occurs in the Big Horn Basin and Granite Mountain in the Wind River Basin to the south. These basins were formed during the Laramide Revolution beginning in late Cretaceous and extending through Eocene time. During Pliocene and Pleistocene, regional uplifting of mountains, faulting and warping, with regional degradation and sediment deposition, occurred to give Wyoming its unique ecology and high elevation (Love 1960).

Soils and topography

Soil classification information is given in Table 1. Topography was generally level to slightly rolling with a general landscape of hills and plains. Slope values range from 0.5 to 5.5%. Exposure of the sites was generally easterly. Elevations ranged from 1,400 to 2,164 m.

Soil profile pits were excavated and characterized at all research locations. Samples from each profile were taken for detailed laboratory

Table 1. Location and site characteristics of two semiarid research sites in western Wyoming.

	Smilo	*Granite Mountain*
County	Washakie	Fremont
Latitude and Longitude	43° 52′ 107° 44′	42° 41′ 108° 07′
TWSP	45 N	31 N
RANGE	91 W	94 W
SECTION	SWSE 14	NESW 7
Soil great group	Natrargid	Haplargid
Soil subgroup	Haplic	Borollic
Soil family (time)	Montmorillonitic, mesic	Loamy, mixed
Soil series	Absted	Forelle
Slope (percent)	5.5	0.5
Exposure (degrees)	076	030
Elevation (meters)	1,400	2,164

Table 2. Detailed physical and chemical soil characteristics of the research sites.

	Smilo	*Granite Mountain*
Depth (cm)		
A Horizon	0–10	0–5
B Horizon	10–28	5–43
C Horizon	28–	43–
Texture (percent)		
Coarse sand	16.2	13.4
Fine sand	22.2	19.6
Silt	31.9	34.5
Clay	29.7	32.5
Textural class	Clay loam	Clay loam
Bulk density A Horizon	1.3	1.1
Bulk density B Horizon	1.7	1.5
Porosity (percent) A Horizon	49	59
Porosity (percent) B Horizon	37	44
pH A Horizon (paste)	7.9	7.6
pH B Horizon (paste)	8.3	7.5
Sol. salts A Horizon (mmhos/cm^2)	0.6	0.7
Sol. Salts B Horizon (mmhos/cm^2)	1.5	0.5
$CaCO_3$ A Horizon (percent)	0.5	0.6
$CaCO_3$ B Horizon (percent)	0.6	10.5

analysis (Ries 1973). Soils were typical Aridisols derived under arid climates and exhibited poorly differentiated horizons (Table 2). Most clay has been leached to lower horizons in the development process. Soluble salt concentrations were low and pH was well within tolerances for growth of most species common to the sagebrush-grass region (Ries 1973, Soil Survey Staff 1967). Variation in horizon thicknesses and textural values occurred which affected soil moisture relationships. The A and B horizons comprised only 28 cm thickness at Smilo but were 43 cm at Granite Mountain. The textural difference of 5.4% less sand at Granite Mountain than at Smilo occurred as a result of almost equivalent increases of silt and clay, 2.6% and 2.8% respectively.

Climate

The climate is continental, characterized by great diurnal and seasonal temperature variation, abundant sunshine, and mean annual precipitation of 12–22 cm. Mean annual temperature for the study sites, derived by extrapolation from nearby US Weather Service Stations, were 11°C at Smilo and 7°C at Granite Mountain. Extremes ranged from 45°C in the Big Horn Basin to a low of −50°C at the Granite Mountain site in the Wind River Basin. On-site recorded seasonal precipitation was greatest during winter and spring (Table 3). Although variable from year to year, summer and fall precipitation occurred consistently, a major climatic variation from that of areas which exhibit Mediterranean moisture

Table 3. Long-term seasonal precipitation and temperature characteristics of the semiarid rangeland research sites, 1960–1982.

	Smilo	*Granite Mountain*
Precipitation (mm)		
Fall	28	28
Winter	71	66
Spring	89	96
Summer	28	33
Annual	216	223
Temperature (centigrade)		
Fall	14	11
Winter	0	−4
Spring	12	12
Summer	22	21
Annual	11	7

distribution with almost complete absence of summer/fall rains, such as occurs on rangelands of Kuwait.

Vegetation

According to Kuchler (1964) the site in the more northerly Big Horn Basin occurred in the wheatgrass-needlegrass-shrub steppe zone (*Agropyron-Stipa-Artemisia*). That in the Wind River Basin occurred in the sagebrush steppe zone (*Artemisia-Agropyron*). The two categories were combined by West (1983) because of the small area of the former and because only inadequate literature had been available to him for differentiation.

Wyoming big sagebrush (*Artemisia tridentata* ssp. *wyomingensis*) and mountain big sagebrush (*A. tridentata* ssp. *vaseyana*) were the dominant shrubs (Appendix 1). Another frequently encountered shrub was Douglas rabbitbrush (*Chrysothamnus viscidiflorus*). The primary perennial grass was western wheatgrass (*Agropyron smithii*); others included Sandberg bluegrass (*Poa secunda*), mutton bluegrass (*P. fendleriana*), prairie junegrass (*Koeleria cristata*), and needle and thread (*Stipa comata*). Additional perennial grass and grass-like species which occurred irregularly were bottlebrush squirreltail (*Sitanion hystrix), blue grama (Bouteloua gracilis*), Indian ricegrass (*Oryzopsis hymenoides*), threadleaf and needleleaf sedge (*Carex filifolia* and *C. eleocharis*). The introduced annual cheatgrass brome (*Bromus tectorum*) on occasion responded dramatically following sagebrush control, but was usually limited by competition from associated perennial species, as also was the native annual sixweeks grass (*Vulpia octoflora*). Common mat-forming and succulent species included Hood's phlox (*Phlox hoodii*), Hooker sandwort (*Arenaria hookeri*), plains pricklypear (*Opuntia polyacantha*) and rose pussytoes (*Antennaria rosea*).

PROCEDURES

Environmental measures

Precipitation gauges were established during the late 1950's. A network of 70 gauges, spaced at 10–20 km intervals, was established initially throughout the Big Horn and Wind River Basins. These gauges were metal cylinders with an orifice diameter of 80 mm. More than 200 gauges have been installed over much of western Wyoming. Precipitation is recorded seasonally (Fisser *et al.* 1966).

A neutron probe was utilized for determination of soil moisture during the 1960's (Fisser *et al.* 1963). Gravimetric sampling with king tubes was conducted at several locations during the 1970's and early 1980's (Fisser *et al.* 1977).

Ambient and time-lapse air temperature changes were recorded with maximum-minimum thermometers. Soil temperatures at 2.5, 20, 38 and 55 cm depths were obtained from permanently placed thermographs (Fisser *et al.* 1963). At sites of phenological studies, soil temperatures were obtained with thermometers inserted into the hole created by the king tube for soil moisture sampling (Fisser *et al.* 1977, Kleinman and Fisser 1976).

Photographic record

Colour and black and white photographs were taken to maintain a permanent visual record of research activities and temporal variation of vegetation and site conditions (Fisser and Ries 1973). In addition to the general photograph showing distant vegetation and skyline of each transect, a close-up was taken to show the first two quadrats. Photographs were taken from one or more corners of each exclosure to provide a visual expression of vegetation inside and outside the fenced units.

For three years during the phenodynamics study, close-up photography of specific sagebrush branches, plains pricklypear clumps, and grass plants was taken at approximately two-week intervals during the growing season. Observation of these segmental records exhibited a surprising characteristic in the production of reproductive stems. A specific branch usually alternated annually, changing from one year with only vegetative growth to the next year with seed production (Kleinman 1976). This was especially evident during years of moisture and temperature stress. Shrubs thus conserved their energy output by restriction of aerial growth to solely vegetative expression, and, in doing so, enhanced root development.

Vegetation monitoring

Area cover and production. Earliest field research efforts initiated in 1961 investigated cover and production of herbaceous species. Basal cover of herbaceous species and crown cover of shrub, half-shrub, succulent and mat-forming species were recorded (Fisser 1961, Fisser and Ries 1971). Area cover and herbage production studies on sagebrush-grass sites were conducted with transects of 20 quadrats, 30 × 30 cm, spaced systematically along a randomly located 30 m steel tape. Area cover percentages of all herbaceous, semi-woody and woody species were estimated within each quadrat. Data of shrub crown cover and basal cover of plains pricklypear (*Opuntia polyacantha*), Hood's phlox (*Phlox hoodii*) and Hooker sandwort (*Arenaria hookeri*) were not combined when comparing area cover to herbaceous production since this group of plants was not clipped. Forage production was determined by clipping herbaceous species at ground or crown level at near-peak standing crop. Exclosures were clipped on or near the same date as the previous years. Clippings were oven-dried at 70°C for 24 hours prior to weighing.

The complete list of plant species recorded at the sites is given in Appendix 1. Several perennial grasses and forbs were common to both sites. The visual aspect of the sites was quite similar but woody plants and most annuals were unique to each location, an obvious expression of adaptation to climatic and other environmental differences.

Shrub production. Beginning in 1971 shrub production was obtained by the modified double-sampling method during September and October. The technique involved a rapid weight-unit estimation procedure. Estimates were made on transects consisting of 20 quadrats, 1.2×1.5 m, spaced from 6–12 steps apart. The number of weight-units in each plot was estimated individually by two men and then checked for accuracy. Weight-units were determined prior to the estimation procedure (Fisser and Ries 1972).

Phenodynamics. To study the relationship of plant phenodynamics with vegetation production and environmental factors, an intensive monitoring program was initiated in 1973 at 12 research locations widely distributed in western Wyoming (Fisser and Kleinman 1974). Phenophase descriptions and stages were established for four primary species: western wheatgrass (*Agropyron smithii*), bluebunch wheatgrass (*A. spicatum*), big sagebrush, black sagebrush (*Artemisia nova*). A general phenology inventory scale was established for use on all other species.

Twenty individual plants of each of the four principal species were permanently located along 30 m lines in the 12 exclosures. At every 1.5 m on the line, directions along and from the line for the closest plant were recorded. Measurements of the shrubs (big sagebrush and black sagebrush) included plant size, age, twig length and seed stalk length. Measurements of the grasses (western wheatgrass and bluebunch wheatgrass) included plant size, age, leaf width and height, spike height, number of spikelets per seed head, and seed head development.

Separate scores were assigned for vegetative and reproductive development because most plant individuals of semiarid regions seldom flower consistently. Time-lapse precipitation accumulation, max–min air temperatures, soil moisture and temperature and photographs were recorded at each observation date (Fisser and Kleinman 1975).

Productivity. Beginning in 1979, plant species and community productivity relationships with temporal and environmental changes were investigated. Early and late-season total population clipping and weight estimate procedures were employed. These were supplemented with multiple period biomass sampling of individual culms and buds of the four prime species (Fisser *et al.* 1980).

Beginning at least by the time of early standing crop estimation, individual plant units of each prime species were clipped. These were selected to best represent the average phenology of a given prime species as indicated from the values of average phenology on the transect lines. The specimens to be clipped were selected at locations near the corresponding phenology transect line. Twenty were clipped and individually bagged with exclosure, species and date identification.

Productivity units of the wheatgrasses were individual culms. Selection was based primarily on the phenology average of the permanent transect lines. Height of culms, although often uniform for most plants at a given stage of development, was specifically removed from the observer selection process.

Productivity units of the sagebrushes were individual twig tip initiation buds and initially included one node from the previous season's development. The inclusion of the added node was a minimal part of the productivity unit weight and was included to simplify the field procedure which required identification of a new bud twig as well as a uniform specified point at which it was to be removed from the shrub. A soft coloured wire was located at the base of the twig and did not affect bud growth development. Phenologic development, representative of the adjacent permanent phenology transect, defined the observer selection criteria. Length of current year twig development was utilized only if it related to identification of phenology which became definitive in some late-season reproductive stages (Fisser *et al.* 1982).

The selection of productivity units at the observation time prior to the clipping date was incorporated into the field procedures because of the intended extrapolation of the prime species productivity data to that of the plant community. Productivity measures of entire plant populations typically involve clipping of all species at frequent intervals throughout the growth season (Dickenson and Dodd 1976). Such data provide estimates of total standing crop at each date. They also required extensive work force and time allocations totally beyond the limits of this study of 12 locations widely distributed over the western two-thirds of Wyoming.

With the data for both early- and mid-season community standing crop, and with productivity data of four prime species at frequent intervals throughout the April–October growing season, extrapolation and interpolation can be utilized to express coefficients of weight-phenology related development of the prime species to that of the plant community populations (Fisser 1984a). Thus, there was an apparent conceptual need, when developing the present field procedures, for time-extended relations of phenology-productivity unit selection and actual harvesting of the individual plant units.

These procedures may not as closely depict true productivity values of community populations as might be determined by clipping of all species. The expression of standing crop at frequent intervals, derived from the productivity weight-units integrated with early- and late-community standing crop, however, does provide a reasonably valid characterization of plant community productivity dynamics (Fisser 1984b). The interrelation of prime species productivity units with phenologic progress provides a detailed data base to describe plant community phenodynamics and related standing crop values (Fisser *et al.* 1983). These data are not available elsewhere and may be utilized for application to other semiarid shrub regions typical of western Wyoming as well as the rangelands of Kuwait.

RESULTS

Regression and time series

For more than 20 years, research activities in Wyoming have involved rangeland production studies and research to develop predictive equations. Research has involved sheep and cattle grazing trials, harvester ant control, wildlife studies, shrub and forb plantings, chemical shrub control, grass establishment, phenology, productivity, premining regulatory classification and many other aspects of semiarid rangeland vegetation, soils and environment. Many permanent quadrats and transect sampling locations have been established to enhance long-term data acquisition. Important findings have been obtained which identify relationships of annual climatic characteristics with plant growth and development rate (Hargis 1980, Kleinman 1976, Noller 1968, Whysong 1973). The statistical approach with various regression equations developed by these authors have shown clearly the definitive relationship between forage yield and seasonal precipitation, primarily winter and spring, and the effects of temperature.

The concept of time-series analysis has recently been developed and utilized in some studies. The importance of this form of analysis relates to the fact that recent data affect the forecast more significantly than the earlier data. Time-variable influences of environmental factors influence both productivity and phenology. By interpretation of cause-effect responses, the biotic and environmental data will provide information useful for prediction of growth and phenodynamics. Thus, future vegetation inventory processes can be utilized to estimate the proportion of probable total crop production at any time during a future growth season, subject to moisture and temperature benefits or restraints.

A parallel to the concept of time-series analysis may be derived from a long-term data set obtained from a Gardner saltbush (*Atriplex gardneri*) community of northcentral Wyoming (Fisser and Joyce 1983). Sheep grazing intensity trials had been initiated on this site in 1955 and continued annually through 1978. Herbage production data were obtained annually and environmental information was compiled with seasonal precipitation data. The temporal vegetation response was investigated in terms of protection from grazing within exclosures constructed in 1956 and response to three levels of winter sheep utilization. Almost total elimination of perennial grasses and forbs had occurred prior to the initiation of the program and reduced vigour of the Gardner saltbush was evident. The vegetation changes during the first seven years from 1955 through 1961 primarily exhibited population modifications suggestive of recovery from many years of continuous overuse. From 1962 through 1971 grass and forb production while minimal, was increasing. Data from the last seven years of 1972 through 1978, however, portrayed dramatically greater grass and forb production and a corresponding decrease in shrub production. By 1978, after 23 years of protection, grass contributed nearly 50% of the total standing crop. In contrast, it contributed less than 1% after seven years of protection in 1961.

The results of this study show that the vegetation composition changed significantly under protection from grazing. Additionally, these changes occurred during a time when seasonal precipitation was also changing. While spring precipitation was decreasing, winter precipitation increased. Analyses of the relationship between precipitation and standing crop, based on the entire period, would have been misleading. It was demonstrated that the changing plant composition affected the relationship of total standing crop response to precipitation. During the earlier period, while the total herbage production was primarily from saltbush, herbage yield response to moisture was minimal. During the later period, however, with recovery from historical overuse essentially complete, and herbaceous species contribution as much as one half or more of the total vegetation production, response to precipitation was much more dynamic. Protection from grazing, then, was only partially responsible for the expression of vegetation trends. An interpretation of climatic and vegetation relations based principally on only early short-term records or only the total 24 year data set, would have suggested a misleading response to non-grazing. By isolating the last seven year data set and comparing that to the first seven, the early recovery from historical overuse could be separated from significant vegetation composition and field changes expressive of modified precipitation occurrence. In addition, the total data set afforded comprehensive study of vegetation changes as well as greater definitive capability for prediction of climatic/management interrelationships (Fisser and Joyce 1983).

This approach could be utilized not only on rangelands similar to the semiarid shrublands of western Wyoming, but very conceivably on rangelands with very different environmental characteristics elsewhere in the world. Kuwait and other areas of the Arabian Gulf possess perennial vegetation complexes (e.g. *Rhanterium/Cyperus*) which exhibit similar annual plant dynamics while responding to a November–May growing season.

Examples abound in the literature expressing the necessity for continuous monitoring of vegetation and environmental data (Green 1984, Joyce 1981). In addition, tremendous numbers of models have been utilized for prediction, most of which have been inadequate. Recent efforts in the Wyoming research program have utilized simple regression relationships of herbage production to precipitation. In most instances, herbage production is usually positively and directly related to winter and spring precipitation in the semiarid regions of western Wyoming. For example, at the Smilo exclosure in the Big Horn Basin of Wyoming, herbage production increased at a rate of 12 kg for each increased cm of winter plus spring precipitation

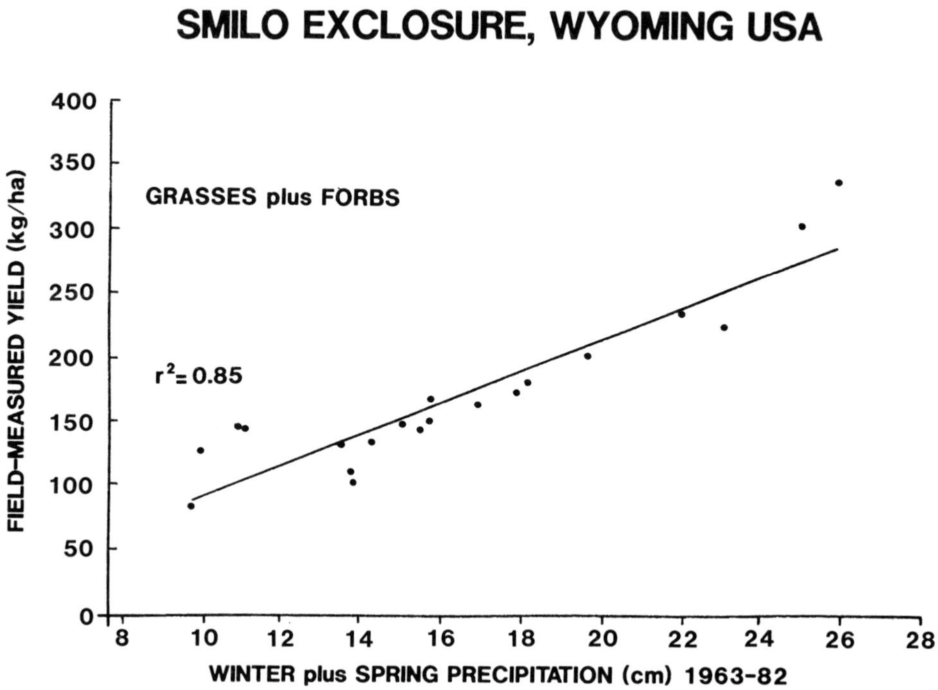

Fig. 2. Comparison of field-measured yield of grasses and forbs and winter plus spring precipitation at Smilo.

($r^2 = 0.85$). These data (Fig. 2), recorded over a 20 year period from 1963–1982, are representative of many sites in semiarid regions, not only of Wyoming but many areas of the western United States and other regions of the world. Studies have also been conducted to determine the relationship of herbage production to precipitation when combining data from several locations in central Wyoming.

Many studies which attempt to show quantitative relationships of herbage yield with precipitation have generally been unsuccessful. This has been due primarily to the variable temporal distribution of precipitation and the fact that range plants respond to soil moisture only during actively growing phenological stages and little, if any, during fall and winter in the northern United States. Use of seasonal, or combinations of monthly, precipitation has helped account for the distribution effects, and in many situations has provided reasonably accurate herbage yield estimates. In Canada, May and June precipitation has provided good estimates of yield. In North Dakota, April through July was closely correlated with annual herbage yield and in the northern Great Plains in the area of eastern Montana and western North Dakota, the May precipitation was the best single month to index herbage production (Joyce 1981).

Since precipitation-herbage yield relationships often exhibit a very site-specific dependency, the development by several researchers has involved the use of ratios of long-term values. A yield and precipitation index model for herbage yield estimation, then, may be adjusted and predictions made with some reasonable expectation of success. The analyses show that the winter plus spring precipitation gave the best overall prediction results. These analyses also suggested that combining the herbage yield from several locations gave most consistent results. The slope of the regression line in this study was approximately 11 kg of herbage production per cm of winter plus spring precipitation (Hanson *et al.* 1984). These data, which included the Granite Mountain exclosure described earlier, obviously were different from those of the Smilo exclosure in the more northerly and xerophytic area of the Big Horn Basin. The indices also were significantly different from research presented by others in other regions of the western United States. These data verify the significance of site-specific differences in herbage yield-climatic data relationships.

Transpiration base modeling

Whether a model can be utilized to predict and forecast vegetation interactions with differing climatic variables through time is dependent upon the quantitative characteristics of the model and of course, on the interpretative evaluation of those conducting the research. With long-term climate records and reasonably long-term data of herbage production, an acceptable estimate of expected forage yield may be developed by those concerned with the ecology of semiarid rangelands.

The generalized overview of modeling which has been approached in our research relates primarily to the water-balance climate model developed by Wight and Hanks (1981). In this model, which is identified by the acronym ERHYM, the herbage yields are determined as a function of the ratio of actual to potential transpiration values (TA/TP). Availability of input data, relative simplicity, and low computer costs make this model a viable tool for both research and resource management.

The ERHYM model was developed in eastern Montana with research conducted on shortgrass rangeland areas. The acronym was taken from the title of the research project name which was Ekalaka Rangeland Hydrology and Yield Model (Wight *et al.* 1984). Data from the Wyoming research sites including the Smilo and Granite Mountain exclosures were utilized as a validation test of the parameters identified from the eastern Montana site.

The major components of the ERHYM model require climatic values of maximum and minimum temperature, daily precipitation and daily solar radiation. Soil information that is utilized includes albedo values of the soils, the water holding capacity, infiltration rates and daily temperature. Daily update of soil water content is estimated by the model as related to the precipitation budget, evapotranspiration and plant growth characteristics of the site. The vegetation components are identified by a crop coefficient and transpiration coefficient, as well as relative root distribution and a relative growth curve factor. Growth is estimated on the basis of the ratio of potential evapotranspiration to actual evapotranspiration. The model integrates several aspects of range environment including amount of precipitation and daily distribution of precipitation. Evaporative demands of the vegetation are also taken into account. Soil moisture by horizon is a major factor and contributes to accurate prediction of rangeland yield.

The Smilo and Granite Mountain exclosures are sagebrush-grass complexes with western wheatgrass as a predominant species. Soil water and soil temperature data for 1963 to 1968 were adequate to develop the model's soil water balance and soil temperature components.

Precipitation was collected in simple aluminium rain gauges installed at the exclosures at each study site. A daily precipitation record was developed by prorating the on-site seasonal precipitation records with nearby US Weather Service Stations using an inverse of the distance-squared procedure. Daily precipitation records obtained in this manner were generally satisfactory for predicting herbage production, but may induce a few day-to-day discrepancies between simulated and field-measured soil water content values.

Daily solar radiation and maximum and minimum air temperatures were obtained from the climate generating sub-model. For comparison purposes, the model also was run using the maximum and minimum air temperatures and actual solar radiation from the US Weather Bureau Station at Lander, Wyoming. Soil water content and soil temperature were measured at the Granite Mountain and Smilo sites periodically during the summer months of 1963 through 1968. Soil water was measured by the neutron scatter method at soil depths of 15, 30, 45, 60, 120 and 150 cm. Soil temperatures were

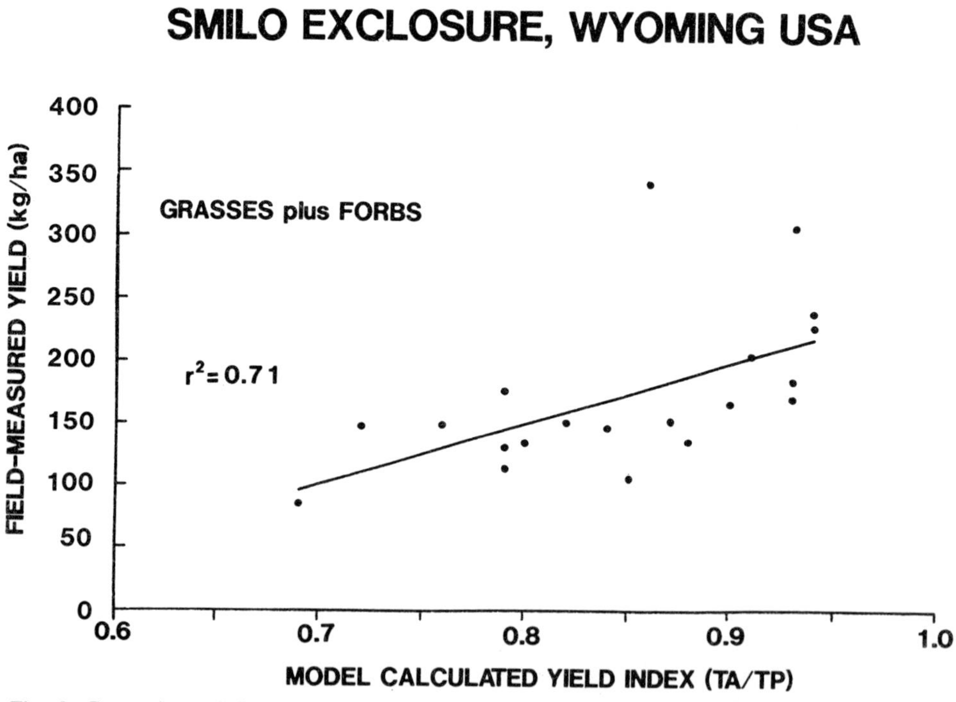

Fig. 3. Comparison of field-measured yield of grasses and forbs and ERHYM model-calculated yield index (TA/TP) of Smilo.

measured at depths of 2, 20, 38 and 56 cm. Herbage yields were determined by clipping herbaceous species at ground or crown level within 30 × 30 cm quadrats.

At the more xeric site with more alkaline soils and reduced infiltration rate, the model-calculated yield index (TA/TP = actual transpiration/potential transpiration) changed by a ratio value of 0.1 with 55 kg/ha change of field-measured yield of grasses plus forbs, $r^2 = 0.71$ (Fig. 3). At the less xerophytic location with reduced soil alkalinity but increased infiltration, the yield index ratio change of 0.1 indicated a change of 85 kg/ha of field-measured yield, $r^2 = 0.78$ (Fig. 4). The 35% greater amount of predicted yield per unit change of the TA/TP ratio was not unexpected.

The effectiveness of ERHYM depends upon its ability to simulate an accurate soil-water balance. Because there is little summer runoff at the sites in Wyoming, the soil-water components of the model were tested using single growing season runs in which actual, field-measured soil water values were used to initialize the model (Fig. 5). Continuous runs were utilized to test the model's ability to predict over-winter recharge. For yield predictions, the model was run in a continuous mode from 1965 through 1980 to match the yield data which were available. Overall, the ERHYM performed reasonably well on the two sagebrush rangeland sites of western Wyoming. Additional validation trials are being conducted to determine the best crop coefficient and transpiration coefficient to improve the accuracy and long-

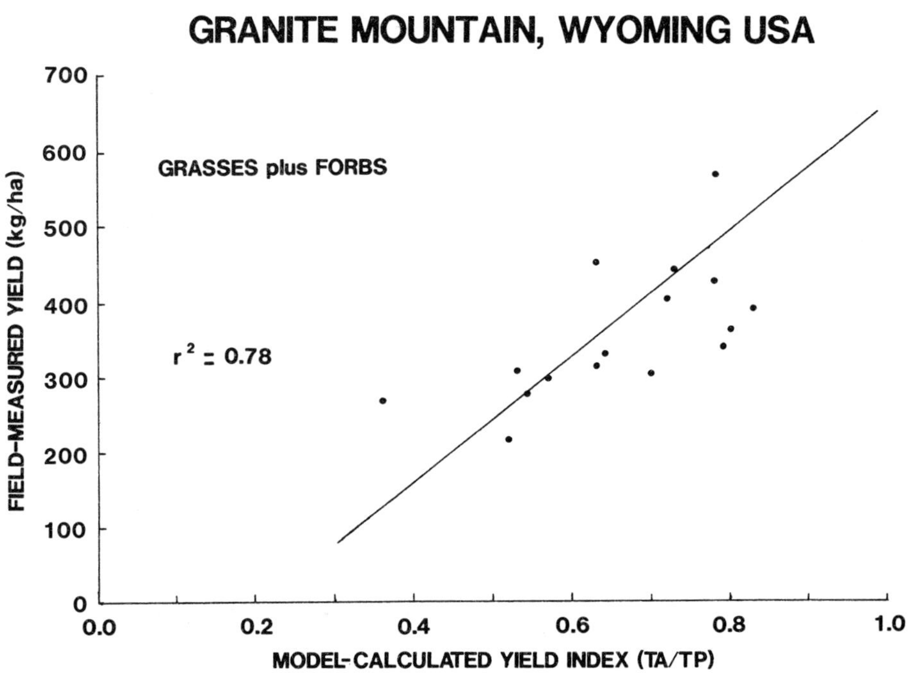

Fig. 4. Comparison of field-measured yield of grasses and forbs and ERHYM model-caculated yield index (TA/TP) of Granite Mountain.

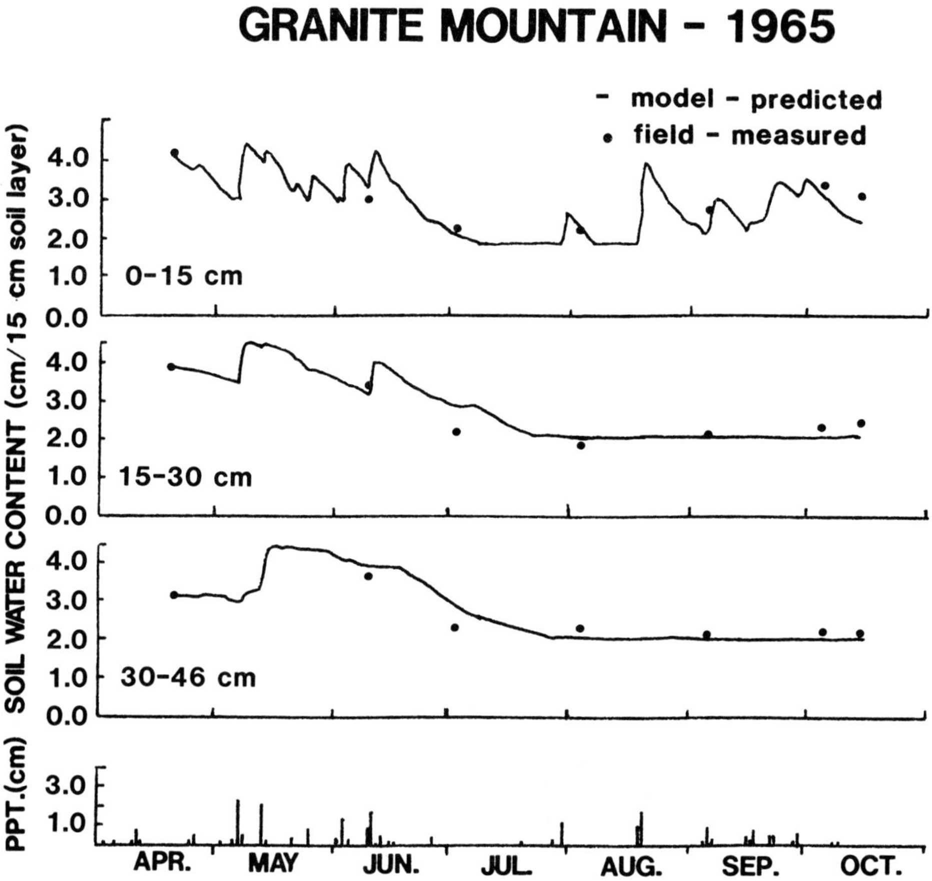

Fig. 5. Comparison of 1965 field-measured and ERHYM model-predicted soil water content at Granite Mountain and daily precipitation (PPT.) during the growing season.

term forecasting ability of the model. The results of this study indicate that the ERHYM model has a wide range of potential applications in research and management of semiarid rangelands.

SUMMARY

The long term research program in Wyoming was developed specifically to generate quantitative and scientific knowledge of rangelands and grazing interaction as well as to provide a data base for application to the often very difficult part of management and use of these lands for grazing animal production.

Grazing studies of *Atriplex* rangelands with winter sheep grazing and summer steer trials exhibited the necessity for controlled intensity of use to maintain and improve depleted ranges. Permanent shrub mapping quadrats are necessary for long-term vegetation analysis because of the typical very slowly changing nature of the perennial shrubs, grasses and forbs of the region. Permanent erosion transects have also exhibited the stability of undisturbed ranges and the drastic instability which may result from specific treatments or practices. The control of harvester ants was beneficial and resulted in forage production increases.

Vegetation inventory for a pre-surface mining activity exhibited the need for strongly controlled management of mined land reclamation activities. A large-scale study of three livestock grazing systems on 30,000 hectares of salt desert shrub rangeland in southern Wyoming was conducted. Precipitation, temperature and soil variations often cause vegetation changes which are much more dramatic than those associated with moderate differences in grazing intensity. Very intensive studies of vegetation phenology were conducted with biweekly growing season sampling on 12 widely distributed sites for 10 years. Interrelated soil moisture and temperature data, aerial temperature and precipitation, significantly modified growth rates and incidence of flowering and fruiting. Separate criteria of vegetation phenologic development were itemized, distinct from that of reproduction and phenologic development, because of the often infrequent flowering and seed production of native desert vegetation.

Indexing procedures were developed to determine individual species productivity and relationships with plant community dynamics. Standing crop of all species was completely sampled twice every year, early and late during the growing season. This was supplemented with timed efficient biweekly single stem and bud clipping of a single grass and a single shrub species. Hence estimates of total plant community productivity at different phenological stages were obtained. These procedures provided a rapid characterization of plant population dynamics not available elsewhere. These may certainly be utilized on semiarid shrub-grass ranges of the world. Models which might be utilized in forecasting impacts of future environmental changes may have significant value. If range managers knew what yields to expect from the rangelands as a result of recent as well as earlier precipitation and temperature variables, they might well be able to avoid or at least minimize drouth impacts by modification of animal numbers and grazing location. Some managers do so on an instinctive basis. Few, however, note with adequate detail previous weather criteria. Forecasting future climatological events, at best, is problematical. With restraints, a model may be developed which will depict generally expected climatic changes.

Virtually hundreds of models have been developed. Few have been able to adequately estimate forage production for long periods. Some have been used to estimate forage production for large regions. Their success has often been dependent on their requiring only easily available climatic data. Statistical regression equations have been utilized for a number of years and have been considered reasonably powerful when compared with those used for estimating variables such as forage production. They have the ability to take into account factors such as vegetation, climate and soils for determination of the factor which most closely approximates the potential cause of relationship of vegetation yield and climatic interactions. They do have a disadvantage when applied in biological systems, however, in that their coefficients remain constant while the factors obviously interact continuously in nature.

For more than 20 years the research activities in Wyoming have involved rangeland production studies and efforts to develop accurate, predictive equations. The regression equations have shown clearly the definitive relationship between forage yield and seasonal precipitation, primarily winter and spring, and the effects of temperature. Recently the concept of time series analysis has been incorporated. This form of analysis addresses patterns of previous relations with more recent data affecting forecasts and predictions more strongly than temporally distant data.

The most recent and comprehensive overview in our approach to modeling relates primarily to the water balance climate model developed by Wight and Hanks. The model is identified by the acronym EHRYM. Required major components are maximum and minimum daily temperature, daily precipitation and daily solar radiation climatic values. Soil albedo and values for water holding capacity and infiltration rates are utilized. Soil water content is updated daily by the model as a function of the precipitation budget, evapotranspiration and the plant growth characteristics of the site. Plant growth is estimated on the basis of actual to potential evapotranspiration. The effectiveness of the model depends on its simulation accuracy with the soil water balance. When compared to field-measured soil moisture and temperature at two sites in Wyoming, it performed reasonably well. Additional validation tests are being conducted to determine best crop coefficient and transpiration coefficient values to improve accuracy and forecasting ability. Results indicate that the ERHYM model has a wide range of potential applications in research and management of semiarid rangelands including those of the Arabian Gulf.

LITERATURE CITED

Beetle, A. A. 1960. A study of sagebrush: The section Tridentatae of *Artemisia. Wyoming Agr. Exp. Sta. Laramie, WY.* Bull. 368.

Beetle, A. A. and K. L. Johnson. 1982. Sagebrush in Wyoming. *Wyoming Agr. Exp. Sta. Laramie, WY.* Bull. 779.

Cook, C. W. and P. L. Sims. 1979. Drought and its relationship to dynamics of primary productivity and production of grazing animals. *In:* Evaluation and mapping of tropical African rangeland; Proceedings of the seminar, 3–8 March, 1975, Bamako, Mali, p. 163–169. International Livestock Centre for Africa, Addis Ababa, Ethiopa.

Daubenmire, R. 1956. Climate as a determinant of vegetation distribution in eastern Washington and northern Oregon. *Ecol. Monog.* **26:** 131–154.

Dickenson, C. E. and J. L. Dodd. 1976. Phenological pattern in the short grass prairie. *Amer. Midl. Nat.* **96 (4):** 367–378.

Fisser, H. G. 1961. Variable plot, square foot plot, and visual estimate for shrub crown cover measurements. *J. Range Manage.* **14 (4):** 202–207.

Fisser, H. G. 1962. An ecological study of *Artemisia tripartita* subsp. *rupicola* and related shrub communities in Wyoming. Ph.D. Dissertation, University of Wyoming, Laramie, WY.

Fisser, H. G. 1963. A study of range conditions in the Big Horn and Wind River Basins, 1962 results, Preliminary report IV. *Wyoming Agr. Exp. Sta. Laramie, WY.*

Fisser, H. G. 1984a. Vegetation productivity and phenology in the Wind River Basin. p. 42–45. *In:* H. G. Fisser and K. L. Johnson, (eds.), Proceedings of the 11th Wyoming Shrub Ecology Workshop. Management of Wyoming Shrublands, Lander, WY. 25–27 May 1982.

Fisser, H. G. 1984b. Productivity dynamics of sagebrush grass ecosystems determined from biseasonal sampling of a single grass and shrub species. *Bull. Ecol. Soc. Amer. Prog. Issu.* **65 (2):** 258.

Fisser, H. G., S. K. Dean, T. L. Deboodt, R. S. Kinucan and J. L. Mulholland. 1982. Phenology and production studies on semiarid shrub-types, 1981 results. *Wyoming Agr. Exp. Sta. Laramie, WY.* Sci. Rep. 1200. Vol. II.

Fisser, H. G., S. K. Dean, T. L. Deboodt, R. S. Kinucan and J. L. Mulholland. 1983. Phenology and production studies on semiarid shrub-types, 1982 results. *Wyoming Agr. Exp. Sta. Laramie, WY.* Sci. Rep. 1230. Vol. I.

Fisser, H. G. and L. A. Joyce. 1983. *Atriplex*/grass and forb relationships under no grazing and shifting precipitation patterns in northcentral Wyoming. p. 87–96. *In:* A. R. Tiedemann, E. D. McArthur, H. C. Stutz, R. Stevens and K. L. Johnson (compilers). Proceedings of the symposium on the biology of *Atriplex* and related chenopods, 2–6 May 1983, Provo. Utah. USDA. Forest Service, *Intermountain Forest and Range Experiment Station.* Gen. Tech. Rep. INT-172.

Fisser, H. G. and L. H. Kleinman. 1974. Phenology and production studies on semiarid shrub-types, 1973 results. *Wyoming Agr. Exp. Sta. Laramie, WY.* Sci. Rep. 607.

Fisser, H. G. and L. H. Kleinman. 1975. Phenology and production studies on semiarid shrub-types, 1974 results. *Wyoming Agr. Exp. Sta. Laramie, WY.* Sci. Rep. 729.

Fisser, H. G., J. T. Nichols and J. R. Wight. 1963. Soil moisture and temperature study, 1963 results, Section IV. *In: Halogeton* Research: Arid land studies of grazing treatments, ecology, shrub improvement and control and moisture relationships. *Wyoming Agr. Exp. Sta. Laramie, WY.* Mimeo.

Fisser, H. G., M. K. Owens, W. J. Waugh and N. E. Hargis. 1980. Phenology and production studies on semiarid shrub-types, 1979 results. *Wyoming Agr. Exp. Sta. Laramie, WY.* Sci. Rep. 1044. Vol. I.

Fisser, H. G. and R. E. Ries. 1971. Arid land ecology research, 1970 results. *Wyoming Agr. Exp. Sta. Laramie, WY.* Sci. Rep. 350.

Fisser, H. G. and R. E. Ries. 1972. Arid land ecology research, 1971 results. *Wyoming Agr. Exp. Sta. Laramie, WY.* Sci. Rep. 428.

Fisser, H. G. and R. E. Ries. 1973. Arid land ecology research, 1972 results. *Wyoming Agr. Exp. Sta. Laramie, WY.* Sci. Rep. 521.

Fisser, H. G., L. R. Robinson, D. R. Kirkham, R. E. Steger and G. L. Noller. 1966. Precipitation pattern study, 1965 results, Section V. *In: Halogeton* Research: Arid land studies of grazing treatments, ecology, shrub improvement and control and moisture relationships. *Wyoming Agr. Exp. Sta. Laramie, WY.* Mimeo.

Fisser, H. G., L. W. Young, J. E. Orpet and N. E. Hargis. 1977. Phenology and production studies on semiarid shrub-types, 1976 results. *Wyoming Agr. Exp. Sta. Laramie, WY.* Sci. Rep. 930.

Goodall, D. W. 1970. Studying the effects of environmental factors on ecosystems. Vol. I, p. 19–28. *In:* D. E. Reichle (ed.), *Ecological studies analysis, and synthesis.* Springer-Verlag, Berlin.

Green, R. H. 1984. Statistical and nonstatistical considerations for environmental monitoring studies. *Environmental Monitoring and Assessment* **4:** 293–301.

Hanson, C. L., H. G. Fisser and J. R. Wight. 1984. Biology and ecology of sagebrush in Wyoming: Herbage yield dynamics and seasonal precipitation relationships. *In:* Proceedings Shrub Research Consortium Third Annual Wildland Shrub Symposium: The biology of *Artemisia* and *Chrysothamnus.* 9–13 July 1984, Provo, Utah, (manuscript in press).

Hargis, N. E. 1980. Phenological comparison of two sagebrush species. M.S. Thesis, University of Wyoming, Laramie.

Heady, H. H. 1984. Sagebrush control and management in the western United States. p. 9–14. *In:* H. G. Fisser and K. L. Johnson, (eds.), Proceedings of the 11th Wyoming Shrub Ecology Workshop, Management of Wyoming Shrublands, Lander, WY. 25–27 May 1982.

Hinds, W. T. and G. M. Van Dyne. 1980. Abiotic subsystems. p. 11–57. *In:* A. I. Breymeyer and G. M. Van Dyne, (eds.), *Grasslands, systems analysis and man.* International Biological Programme 19. Cambridge University Press, Cambridge, England.

Johnson, K. L. 1979. Basic synecological relationships of the sagebrush types on the high plains of Montana, Wyoming and the Dakotas. p. 237–251. *In:* G. F. Gifford, F. E. Busby and J. P. Shaw, (eds.), *The sagebrush ecosystem: a symposium.* Utah State University, College of Natural Resources, Logan, Utah.

Johnson, W. M. 1969. Life expectancy of a sagebrush control in central Wyoming. *J. Range Manage.* **22 (3):** 177–182.

Joyce, L. A. 1981. Climate/vegetation relations in the Northern Great Plains and the Wyoming North Central Basins. Ph.D. Dissertation, Colorado State University, Fort Collins.

Kearl, W. G. 1980. Multiple uses and abuses of public lands. University of Wyoming, Div. of Agr. Econ. Laramie, WY. AE80-17R.

Kearl, W. G. and J. W. Freeburn. 1980. Economics of big sagebrush control for mitigating reductions of federal grazing permits. University of Wyoming, Div. of Agr. Econ. Laramie, WY. AE80-05R.

Kinucan, R. J. and H. G. Fisser. 1984. Dynamics of vegetative plant part phenodynamics and related stress adaptations. *Bull. Ecol. Soc. Amer. Pro. Issu.* **65 (2):** 253.

Kleinman, L. H. 1976. Phenodynamics and ecology of sagebrush-grass rangelands. Ph.D. Dissertation, University of Wyoming, Laramie.

Kleinman, L. H. and H. G. Fisser. 1976. Phenology and environmental relationships of sagebrush-grass rangelands. *J. Colo.-Wyo. Acad. Sci.* **8 (1):** 48–49.

Kuchler, A. W. 1964. *Potential natural vegetation of the conterminous United States.* Amer. Geog. Soc., New York, NY. Spec. Publ. 36.

Levitt, J. 1964. Drought. p. 57–66. *In: Forage plant physiology and soil range relationships.* Amer. Soc. Agron. ASA Special Pub. No. 5.

Lieth, H. 1970. Phenology in productivity studies. p. 29–24. *In:* D. E. Reichle, (ed.), *Analysis of temperate forest ecosystems.* Springer-Verlag, Berlin.

Love, J. E. 1960. Cenozoic sedimentation and crustal movement in Wyoming. *Amer. J. Sci.* **258-A:** 204–214.

Moss, D. N. 1964. Some aspects of microclimatology important in forage plant physiology. p. 1–14. *In: Forage plant physiology and soil range relationships.* Amer. Soc. Agron. ASA Spec. Publ. No. 5.

Nelson, A. 1898. The Red Desert of Wyoming and its forage resources. USDA, Div. of Agrostology, Laramie, WY. Bull. 13.

Noller, G. L. 1968. The relationship of forage production to precipitation, cover and soils in northcentral Wyoming. Ph.D. Dissertation, University of Wyoming, Laramie.

Olson, R. A. and W. D. Gerhart. 1982. A physical and biological characterization of riparian habitat and its importance to wildlife in Wyoming. Wyoming Game and Fish Dept. Cheyenne, WY.

Owen, D. F. 1979. Drought and desertification in Africa: Lessons from the Nairobi Conference. *Oikos* **33:** 189–191.

Passey, H. G. and V. K. Hugie. 1962. Sagebrush on relic ranges in the Snake River Plains and northern Great Basin. *J. Range Manage.* **15 (5):** 273–278.

Porter, C. L. 1960. A flora of Wyoming, Part I. *Wyoming Agr. Exp. Sta. Laramie, WY.* Bull. 402.

Ries, R. E. 1973. Quantitative evaluation of herbage production and site factors on ecologically diverse sagebrush–grass locations in Wyoming. Ph.D. Dissertation, University of Wyoming, Laramie, WY.

Robertson, J. H. 1947. Responses of range grasses to different intensities of competition with sagebrush (*Artemisia tridentata* Nutt.). *Ecology.* **28 (1):** 1–16.

Sneva, F. A. and C. M. Britton. 1983. Adjusting and forecasting herbage yields in the intermountain big sagebrush region of the Steppe province. *Oregon Agr. Exp. Sta. Corvallis, OR.* Bull. 659.

Soil Survey Staff. 1967. Supplement to soil classification system, 7th Approximation. USDA, Soil Conservation Service. Washington, D.C.

Stoddard, C. A., A. D. Smith and T. W. Box. 1975. *Range management.* 3rd ed. McGraw-Hill, New York.

Vale, T. R. 1975. Presettlement vegetation in the sagebrush-grass area of intermountain west. *J. Range Manage.* **28 (1):** 32–36.

West, N. E. 1983. *Ecosystems of the world 5:* Temperate deserts and semi-deserts. Elsevier Scientific Publishing Company, Amsterdam, The Netherlands.

Whitman, W. C. and C. N. Haugse. 1972. Effects of added rainfall on range vegetation production. *North Dakota Agri. Exp. Sta. Fargo, ND.* Progress Report.

Wight, J. R., H. G. Fisher and C. L. Hanson. 1984. Biology and ecology of sagebrush in Wyoming: Validation of a rangeland production model (ERHYM) for sagebrush sites. *In:* Proceedings Shrub Research Consortium Third Annual Wildland Shrub Symposium: The biology of *Artemisia* and *Chrysothamnus.* 9–13 July 1984, Provo, Utah, (manuscript in press).

Wight, J. R. and R. J. Hanks. 1981. A water-balance climate model for range herbage production. *J. Range Manage.* **34:** 307–311.

Whysong, H. L. 1973. Development and evaluation of statistical models for predicting perennial grass production in northcentral Wyoming. Ph.D. Dissertation, University of Wyoming, Laramie.

Appendix 1. List of plants which occurred on the study sites.

		Smilo	*Granite Mountain*
Shrubs, half-shrubs, mat-form, succulent			
Machaeranthera glabriuscula	Half-shrub	X	
Opuntia polyacantha	Succulent	X	
Arenaria hookeri	Mat-form	X	
Artemisia tridentata ssp. *vaseyana*	Shrub	X	
Artemisia tridentata spp. *wyomingensis*	Shrub		X
Antennaria rosea	Mat-form		X
Chrysothamnus viscidiflorus	Shrub		X
Eriogonum ovalifolium	Mat-form		X
Haplopappus acaulis	Mat-form		X
Phlox hoodii	Mat-form		X
Perennial grasses and grass-like			
Bouteloua gracilis	*Grass*	X	
Carex filifolia	Sedge	X	
Agropyron smithii	Grass	X	X
Poa secunda	Grass	X	X
Sitanion hystrix	Grass	X	X
Stipa comata	Grass	X	X
Carex eleocharis	Sedge		X
Koeleria cristata	Grass		X
Oryzopsis hymenoides	Grass		X
Poa fendleriana	Grass		X
Perennial forbs			
Castilleja angustifolia		X	
Crepis occidentalis		X	
Cryptantha bradburiana		X	
Cymopterus montanus		X	
Lomatium orientale		X	
Penstemon cleburnei		X	
Sphaeralcea coccinea		X	
Allium textile		X	X
Astragalus purshii		X	X
Erigeron pumilus		X	X
Machaeranthera canescens		X	X
Agoseris glauca			X
Arabis holboellii			X
Astragalus miser			X
Astragalus missouriensis			X
Comandra pallida			X
Crepis modocensis			X
Erigeron compositus			X
Erysimum asperum			X
Lesquerella ludoviciana			X
Lithophragma parviflora			X
Lithospermum ruderale			X
Penstemon laricifolius			X
Penstemon glaber			X
Senecio canus			X
Trifolium gymnocarpon			X

Appendix 1. (cont.)

		Granite	
		Smilo	*Mountain*
Annuals			
Androsace septentrionalis	Forb	X	
Bromus tectorum	Grass	X	
Chenopodium leptophyllum	Forb	X	
Descurainia pinnata	Forb	X	
Gilia pumila	Forb	X	
Lepidium densiflorum	Forb	X	
Machaeranthera tanacetifolia	Forb	X	
Plantago patagonica	Forb	X	
Plantago spinescens	Forb	X	
Salsola kali	Forb	X	
Vulpia octoflora	Grass	X	
Lupinus pusillus	Forb	X	
Lappula redowskii	Forb	X	X
Sisymbrium altissimum	Forb	X	X
Bromus japonicus	Grass		X
Camelina microcarpa	Forb		X
Cordylanthus ramosus	Forb		X
Eriogonum cernum	Forb		X
Gymnosteris parvula	Forb		X
Oenothera contorta	Forb		X

Rangeland ecology of the Arab Gulf countries

K. H. BATANOUNY†

Department of Botany, Faculty of Science, University of Cairo, Giza, Egypt

ABSTRACT

The surveyed territory, with an area of 2,560,074 km^2 in six countries, constitutes the major part of the immemorial grazing land of Arabia. Rangelands occupy 972,444 km^2, of which 30% is in a seriously depleted condition. Basic ecological studies, necessary for range management, are lacking. Even the available fragmentary data have been obtained by different methodologies with various concepts.

The physiographic features and the rain peculiarities are the main factors affecting the plant life in the study area which exhibits a remarkable diversity of landforms. The local topography, through its integrated effect on water resources, soil properties and other environmental factors, affects the distribution, establishment, growth and regeneration of plants.

Rainfall data at 24 stations evince that the average annual rains are scanty, ranging from 22.6 to 116.8 mm in the study territory, with the exception of the western highlands and Oman mountains. In addition to aridity, rainfall is variable both in time and space. This reduces the number of options to range management.

Numerous plant communities occur in the study area. Their number would be related to the diverse sets of environmental conditions which occur in this vast area. Only the plant communities which contribute to the range potentiality of the studied rangelands are described. These communities comprise desert, halophytic and mountainous plant communities. The geographical distribution, features of the habitats, the associated species and the range importance of these communities are given. Secondary plant communities replacing some of these communities due to degradation are mentioned.

Consideration of the ecological factors and processes in range management, on both the species and community levels, has been discussed.

LAND AND POPULATION

The territory under study, in terms of its political units, includes: Bahrain, Kuwait, Oman, Qatar, Saudi Arabia and the United Arab Emirates

†Present address: Department of Botany, Faculty of Science, Cairo University, Giza, Egypt.

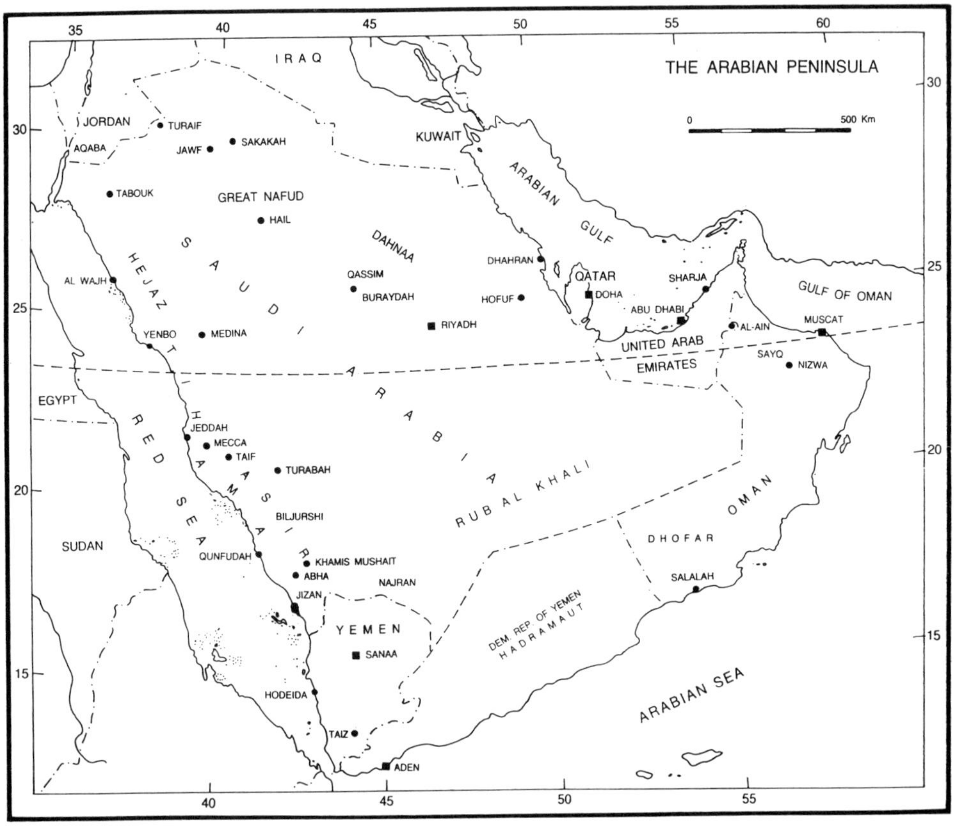

Fig. 1. Location of the Arabian Peninsula.

(Fig. 1). It occupies an area of 2,560,074 km^2, which is about 84% of the total area of the Arabian Peninsula and some 2% of the earth surface of the globe. The largest country, Saudi Arabia, occupies 87.5% of the study territory, while the smallest, Bahrain, only 0.02%, being represented by small islands in the Arabian Gulf. It is to be noted that estimates of area vary in different sources. The same holds true for population and land-use figures. Crosschecks were made in the consulted references from which these data are collated.

The total estimated population of the study territory was 8,375,000 inhabitants in 1965, which increased progressively, reaching 13,837,000 inhabitants in 1980 (Table 1). This increase is referred to the normal rates of population increase and the considerable immigrant population from various countries. The highest density of population is in Bahrain (536/km^2) while the lowest is in Saudi Arabia (4.4/km^2).

One should take into consideration that there are vast areas, especially in Saudi Arabia and Oman, which are almost uninhabited, e.g. the Nafud, the Dahnaa and the Rub' Al-Khali. On the other hand, high population density is remarkable in many cities of the study territory. It is noteworthy that the densest rural population is in the southern Tihama area.

Table 1. Surface area and population of the Arab Gulf countries.

Country	*Area (km^2)*	*Population (1000)* 1965	1970	1975	1980
Bahrain	669	189	216	260	359
Kuwait	17,818	467	738	994	1,356
Oman	212,460	565	750	769	1,500
Qatar	11,427	68	111	180	260
Saudi Arabia	2,240,000	6,975	7,740	8,962	9,319
United Arab Emirates	77,700	111	226	656	1,043
Total	2,560,074	8,375	9,781	11,821	13,837

The rapid changes, since oil exploitation, have been noticeable amongst Bedouins and the urban population. There has been a remarkable drift of manpower from the rural areas to the cities and oil producing centres in quest of a better standard of living and better housing and accommodation. The demand for labour of all kinds has drawn an enormous immigrant population from all the corners of the world.

Data on the nomad population are not available. However, El-Khatib (1974) estimates the Bedouins in Saudi Arabia to be about 14% of the total population. Lack of such data is among the reasons that hamper the development of the rangeland in this territory.

LAND-USE PATTERN

It is extremely difficult to obtain reliable data on the various aspects of land-use in the study territory. Different sources have been consulted (*cf.* FAO 1973 and 1974, Anonym. 1973, Statistical Year Book, Oman 1978 and 1980, AOAD 1981 and 1983, Annual Statistical Abstract, Qatar 1984). The collated data are given in Table 2. The cultivated area under irrigation in the study territory is estimated to be 4711 km^2, representing only 0.18% of the total area. The best estimates for arable land do not exceed 3%. The contribution of agriculture to the national economy of the countries in this territory has been very low. These countries have become increasingly dependent on imports of food and other agricultural products. The settled agricultural communities in most cases have not yet advanced far enough from a subsistence economy. Lack of water resources is among the obstacles and represents the main constraint hindering the use of arable lands for cultivation.

Forests and scrubland, mainly in the highlands, occupy less than 1% of the study territory. They are subject to severe deterioration in many parts. Afforestation trials are practised in all the countries.

The rangelands in the study territory occupy an area of 972,444 km^2 or 38% of the total area under study. Large parts of the grazing areas of the rangelands are either already overgrazed or are gradually deteriorating.

Table 2. Land-use in the Arab Gulf countries (N.A. = data not available).

Country	Area (km^2)	Rangeland (km^2)	Forests, woodland including man-made (km^2)	Arable land (km^2)		
				Rainfed	Irrigated	Uncultivated
Bahrain	669	40	—	—	37	—
Kuwait	17,818	134	22	—	13	159
Oman	212,460	100,000	N.A.	N.A.	411	N.A.
Qatar	11,427	500	4	—	36	53
Saudi Arabia	2,240,000	869,770	16,010	710	3,950	N.A.
United Arab Emirates	77,700	2,000	291	—	264	N.A.
Total	2,560,074	972,444	—	—	4,711	—
%	100	38*	—	—	0.18	—

* FAO (1979) gives estimates of rangelands of these countries as ranging between 96.8 and 99.8% of the total land area.

The existing vegetative cover is the result of climatic and edaphic factors as well as of human and grazing factors. Surveys in Saudi Arabia (El-Khatib 1974) have shown that over 30% of the grazing land is in a depleted condition due to unrestricted grazing, destructive gathering of wood and shrubs and dry farming. Construction activities, especially of roads, resulted in the removal of considerable amounts of plants.

Despite the degradation of soil and plant cover in many parts, there are considerable prospects for improvement. Relic areas of ungrazed vegetation suggest that vegetative cover in the past few decades was more dense than it is today. In fenced areas, the plants exhibit prosperous growth and palatable plants grow vigorously. The remnants of good perennial forage which exist in some localities in the study territory and which have resisted ages of misuse, are able to regenerate the grazing resources, provided prudent grazing practices are used.

PHYSIOGRAPHY AND PLANT LIFE

Several main physiographic units can be distinguished in the surveyed territory. These include:

1. The coastal zones, including those of the Arabian Gulf, Gulf of Oman, Arabian Sea and Red Sea.
2. Nejd and central Arabia, including the escarpment and the plateau regions.
3. The northern region of Arabia.
4. The great sandy regions, including the Nafud, Ad-Dahnaa and the Rub' Al-Khali.
5. The mountains, including the western highlands, the Oman mountains and Dhofar mountains.

These major units vary widely as regards the following criteria:

(a) geographical location (latitude), (b) proximity to water bodies, (c) rainfall regime, (d) geological formations, (e) soil characteristics, and (f) water resources.

The variation in such environmental factors is reflected on the plant life. Hence, different vegetation types occur in these units, which differ in their floristic composition, plant life forms and range potentiality. Despite the oversimplification of the above classification of the main physiographic units, one should take into consideration that each unit exhibits a remarkable diversity of landscapes.

TOPOGRAPHY AND WATER RESOURCES

Values of the measured rainfall in vast areas of the surveyed territory are too low to support a perennial vegetational cover. Yet through runoff, which is influenced by local topography, enough water may collect in certain locations to allow a reasonable vegetation cover. Walter (1963) writes: "If the runoff in a region with 25 mm precipitation is 80%, and the area in which the water accumulates constitutes 4% of the total area, that area will receive an amount of water corresponding to a rainfall of approximately 500 mm." This evinces how much water could be received by the main channels of a wadi. Where there are no wadis, accumulation of sheet runoff in depressions is of great importance in enriching the water revenue of the depressions.

The type of vegetation coincides with the pattern of water resources of the habitat. Three main types are distinguished: (a) accidental (Kassas 1966) in the rainless parts where rainfall is not an annually recurring incident, e.g. in the Rub' Al-Khali; (b) restricted type (Walter 1963) or mode contracté (Monod 1954) in arid areas where rainfall, though low and variable is an annually recurring phenomenon; vegetation is confined to rather restricted areas (wadis, runnels and depressions with adequate water supply; (c) diffuse type in less arid areas, which is a vegetation that is more or less evenly distributed. The latter type occurs in the mountainous regions with relatively considerable rainfall. The second type (restricted type) is the result of runoff and accumulation of water in low parts.

TOPOGRAPHY AND SOIL PROPERTIES

Soil properties, especially the physical attributes, are controlled by topographic features and by physical agents of transport. Low parts are collecting areas which receive runoff water and waterborne and wind-blown sediments. On the other hand, elevated parts have limited water resources and shallow soil, which is usually more coarse-textured than that of lower parts.

Soil texture is among the factors affecting the amount of water that a soil profile can store as well as the amount of water which is available to the plant. Germination of seeds and herbage growth are more responsive to small rainfall events on sandy soil than on clay soil. After rains, the water potential at the wetting front in a dry sand may remain higher than conventional field capacity, so seeds in the wet soil above the front will absorb water and germinate rapidly (Carbon 1975). This is due to the very low hydraulic conductivity of dry sands (Johns *et al.* 1984).

CLIMATE

The climate of the surveyed territory is not uniform due to the extensive land area, proximity to or remoteness from water bodies, and topographic variations. According to its geographical position and vegetation, Arabia

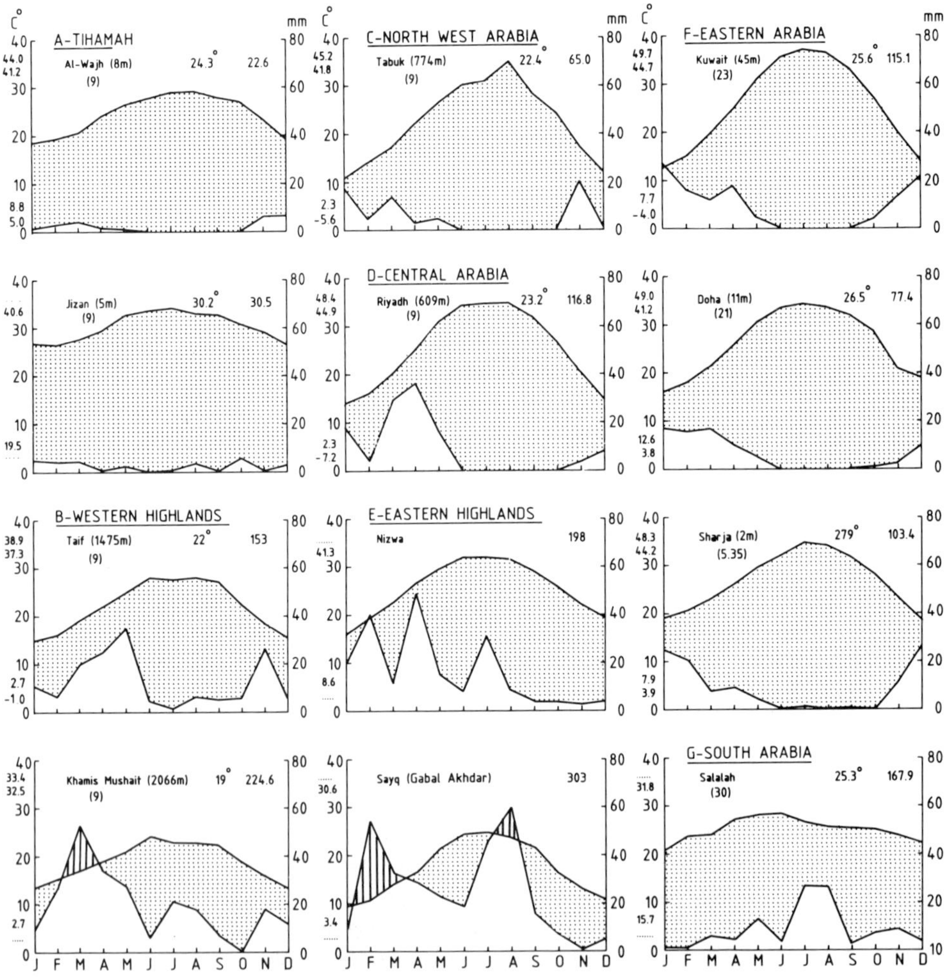

CLIMATIC DIAGRAMS OF SOME STATIONS IN THE ARAB GULF COUNTRIES

Fig. 2. Climatic diagrams of some stations in the Arabian Peninsula.

can be divided into a tropical and an extratropical province (Zohary 1973). The former is governed by two climatic variants: Nubian and Ethiopian. The Nubian variant is extremely desertic, prevalent all along the lowlands bordering on the Red Sea and the Gulf, while the Ethiopian dominates the highlands. The northern part of Arabia is governed by winter rains which in some places are not negligible, although they may be sporadic. Rain-bearing cyclonic depressions which originate in the Atlantic or Mediterranean may move southwards across the peninsula during winter months.

The Arabian Peninsula comprises arid and extremely arid zones, except for the highlands in the southwest and a small area bordering the Gulf of Oman (Meigs 1953). Extremely arid zones comprise the northern part of the Red Sea, the southwestern and the southeastern coasts of the peninsula and the Rub' Al-Khali.

The climatic diagrams (Walter 1955) (Fig. 2) show that with the exception of the highlands in the southwest and Jabal Al-Akhdar in Oman, the area belongs to the subtropical dry zone of the deserts. Rainfall is scanty in the whole peninsula and rarely exceeds an annual mean of 100 mm (Table 3). Rain falls mostly in winter, but it exhibits a change in seasonality as one proceeds to southern latitudes, becoming mainly summer rains in Salalah.

With respect to the mountainous regions, rainfall is relatively high and occurs throughout the year. It is to be noted that summer rains increase as one proceeds southwards, with a maximum of 287.0 mm at Sayq in Jebel Al-Akhdar (Oman).

Fig. 2 and Table 3 show that rainfall in the study territory is scanty. Moreover, it is characterized by irregularity and variability, both in time and space. One of the main characteristics of rainfall in the study area is the occurrence of heavy, sporadic rains of a local nature, which usually cause severe torrents. During these storms the rainfall on one day may exceed the average annual rainfall at the same station.

Air temperature is high in summer, reaching an annual mean of 30.5°C at Jizan. The effect of altitude on temperature is evinced from the low annual mean temperature at Khamis Mushait, being 19°C. Daily and seasonal variations are wide. In coastal areas, maritime influences restrict the temperature range, but this effect disappears a short distance inland. July is generally the hottest month, whereas January is the coldest month.

Freezing temperatures are not infrequent in northern Arabia. Absolute minimum temperature of −5.6°C has been recorded in Tabuk. Records of air temperature ranging from 45 to 50°C are to be expected. An absolute maximum of 51.7°C has been recorded in Abqaiq (25° 56′ N, 54° 28′ E) in the eastern zone of the study area.

There is a considerable local variation in air humidity. Whilst in general the average is low, due to the prevalence of open desert, the coastal regions have remarkably high humidities and dew condensation is common in these regions.

Potential evapotranspiration is very high in the major part of the peninsula. In the northern part of the study area at Rafha, the annual mean evapotranspiration amounts to 1.93 m, while it reaches 2.27 m in the central

Table 3. Average monthly and annual rainfall at different stations in the Arab Gulf countries.

Station	Lat. N ° ′	Long. E ° ′	Alt. (m)	Rainfall (mm) Jan	Feb	Mar	Apr	May	Jun	Jul	Aug	Sept	Oct	Nov	Dec	Mean Annual (mm)
I – *ARABIAN GULF ZONE*																
(a) Coastal Stations																
1. Kuwait	29 13	47 58	45	27.8	15.9	12.1	17.6	4.7	Tr.	—	—	Tr.	3.5	12.7	20.8	115.1
2. Dhahran	26 17	50 09	22	9.7	15.3	8.4	17.2	0.4	—	—	—	—	2.0	3.0	12.1	68.1
3. Doha			11	17.0	15.7	16.6	9.6	5.0	—	Tr.	Tr.	—	1.0	2.3	10.2	77.4
(b) Inland Stations																
4. Hofuf	25 30	49 34	160	9.4	16.2	4.0	7.2	1.6	—	—	—	—	—	3.2	17.6	59.2
5. Al-Ain				10.2	27.8	16.6	7.0	0.3	1.9	2.5	4.5	3.3	—	1.9	1.8	77.8
II – *GULF OF OMAN ZONE*																
6. Sohar				27.9	16.8	8.8	11.0	5.7	0.6	5.0	3.7	—	—	8.5	18.3	106.3
III – *ARABIAN SEA ZONE*																
7. Salalah			18	1.0	1.0	5.0	9.0	13.0	7.0	25.0	24.0	3.0	7.0	7.0	3.0	105.0
IV – *MOUNTAINS IN OMAN*																
8. Nizwa				19.0	40.0	12.0	49.0	15.0	8.0	31.0	9.0	4.0	4.0	3.0	4.0	198.0
9. Sayq				10.0	54.0	33.0	29.0	23.0	19.0	46.0	60.0	16.0	7.0	1.0	5.0	257.0
V – *NORTHERN ARABIA*																
10. Tabuk	28 24	36 35	774	17.3	4.7	13.6	2.7	4.8	—	—	—	—	—	20.4	1.5	65.0
11. Turaif	31 41	38 39	829	8.4	1.3	4.2	5.8	0.9	—	—	—	—	1.0	1.2	6.2	29.0
12. Sakakah	29 58	40 12	574	7.4	3.2	4.4	12.9	0.1	—	0.5	—	—	5.4	27.6	4.2	69.7
VI – *CENTRAL ARABIA*																
13. Qassim	26 18	43 58	650	17.4	10.2	27.0	16.0	10.2	—	—	—	—	—	1.6	7.5	106.9
14. Riyadh	24 42	46 43	609	17.9	4.5	29.4	36.3	16.4	—	0.4	Tr.	—	—	3.8	7.7	116.8
VII – *HEJAZ PLATEAU*																
15. Medina	24 39	39 39	672	13.6	1.0	9.2	10.0	5.0	0.4	—	—	—	0.7	5.3	2.0	47.2
16. Turabah	21 13	41 39	1,130	16.2	4.6	15.1	27.1	10.2	0.9	—	—	3.5	0.2	19.5	1.1	98.4
17. Najran	17 29	44 09	1,150	19.4	3.9	3.0	8.6	2.8	0.1	1.4	1.0	0.9	—	2.0	—	43.1
IIX – *SOUTH WESTERN MOUNTAINS*																
18. Taif	21 29	40 32	1,475	10.4	6.4	19.8	24.8	34.8	4.5	1.4	6.3	5.2	6.0	26.8	6.6	153.0
19. Khamis Mushait	18 18	42 48	2,066	9.1	16.8	52.9	34.2	27.6	6.4	21.6	18.1	7.4	0.4	18.1	12.0	224.6
20. Biljurshi	19 51	41 34	2,400	52.7	36.7	13.6	45.7	22.9	6.7	15.4	17.6	6.6	19.6	76.9	31.9	346.3
21. Abha	18 13	42 30	2,190	43.1	20.4	43.4	77.4	32.3	13.5	50.9	46.6	2.0	2.5	39.4	15.4	386.9
IX – *RED SEA COASTAL ZONE*																
22. Al-Wajh	26 13	36 27	8	1.2	2.4	4.1	1.6	0.8	—	—	—	—	—	6.2	6.3	22.6
23. Jeddah	21 30	39 12	11	26.3	11.0	0.4	10.4	0.2	—	—	—	—	—	21.6	9.9	79.8
24. Jizan	17 03	42 57	5	4.9	4.2	4.7	0.9	2.7	—	0.7	3.6	0.1	5.7	0.3	0.2	30.5

zone at Harad. At Al-Ain in the U.A.E., values of 2.034 m/year were recorded.

VEGETATION

Vegetation studies in the Arab Gulf countries are fragmentary and are mainly reconnaissance surveys. In the late fifties and early sixties, locust expeditions in Arabia resulted in some reports dealing with the flora, vegetation and environment in the region (Popov and Zeller 1963). Vesey-Fitzgerald (1955, 1957a and b) gives useful information on the vegetation of the Red Sea coast, central and eastern Arabia. International Consultant Firms (ITALCONSUL 1969a and b, SOGREAH 1968, 1970 and Parsons 1968) carried out grazing resources inventory in different parts of Saudia Arabia. Information on the vegetation of the different Arab Gulf countries is given in reports prepared by missions and experts of International Organizations (Heady 1963, Kernick 1966). Giacomini and De Marco (1975) studied the vegetation of some parts of the eastern zone of Saudi Arabia. The vegetation of Kuwait has been studied by Halwagy and Halwagy (1974). Vegetation studies in the Hejaz mountains were undertaken by Batanouny (1979) and Batanouny and Baeshin (1983). The vegetation of the Peninsula of Qatar has been investigated by Batanouny (1981) and Batanouny and Turki (1983). A reconnaissance investigation of the vegetation of Oman has been given by Mandaville (1977). However, the available data and accounts of the vegetation of the Arab Gulf countries are scanty and the vegetation in the surveyed territory awaits further studies.

In the present investigation, it is attempted to compile the available information in addition to the records obtained by the author during his excursions and botanical studies in Arab Gulf countries. The main plant communities with reasonable range potentiality will be considered. These communities are either dominated by key range plants or have associates contributing to the range potentiality of these communities. It should be noted that this is not a complete census of the plant communities in this vast area.

The communities given in this study could be classified into three main groups, namely, (a) desert plant communities, dominated by grasses, sedges, undershrubs (succulent or non-succulent), shrubs and trees; (b) halophytic plant communities: inland and coastal salines; and (c) mountainous plant communities dominated by trees occurring at altitudes above 1,000 m.

I. DESERT PLANT COMMUNITIES

A. Communities dominated by grasses and sedges

1. *Panicum turgidum* community

Panicum turgidum Forssk. (Arabic: Tumam, Thumam) is a widespread perennial, sand-binding grass. The plant forms phytogenic mounds of considerable size and grows in dense bushes up to 1 m high or more. The culms are woody, densely branched, with clusters of empty leaf sheaths at the swollen nodes. The grass grows annually both from buds at the base of the stem and from axillary buds on the previous year's stalks. It is generally eaten in the green state by all animals. When dry, its palatability diminishes and only the fine tips are browsed by sheep; however, camels may still consume virtually all parts.

Panicum is a drought-resistant C_4 plant (Ziegler *et al.* 1981) which exhibits high growth rates in late spring as a result of the rise in air temperature. It has a wide geographical amplitude, dominating a widespread community in the surveyed territory and is an important associate in other plant communities. Associates vary widely in the different regions where this community occurs. The common associates include: *Lasiurus hirsutus*, *Pennisetum divisum*, *Calligonum comosum*, *Cyperus conglomeratus*, *Moltkiopsis ciliata*, *Asthenatherum forsskalii*, *Fagonia* spp. and *Monsonia nivea*. In some habitats, the tree and shrub layer is represented by a few individuals of *Acacia tortilis* and *Lycium shawii*. On relatively fine-textured soils in wadis, associates such as *Hammada elegans* are present, while in habitats subject to water erosion, there are associates like *Rhazya stricta* and *Aerva javanica*. In the Red Sea coastal zone, *Dipterygium glaucum*, *Zygophyllum* spp. and *Abutilon pannosum* are common associates.

In the rainy season, numerous annual species grow in the spaces between the perennial framework. These include: *Schismus arabicus*, *Trigonella* spp., *Plantago* spp., *Launaea* spp., *Horwoodia dicksoniae*, *Silene villosa* and many others. These ephemerals contribute to a considerable extent to the range potential of this community. The plant cover in summer may range from 5 to 20%, but it increases in rainy years in spring to 40% or even more.

Degradation of the habitat supporting the *Panicum* community results in the appearance of non- or less-palatable species. The secondary vegetation varies in the different regions and depends on the land-use pattern. Removal of *Panicum* for cultivation of millet in the Tihama region or through overgrazing results in the appearance of a secondary plant growth dominated by *Dipterygium glaucum* on fallow or disturbed land. In Hejaz and central Arabia, overgrazing and consequently the degradation of the habitat supporting the *Panicum* community results in the appearance of *Rhazya stricta*. Degraded habitats of the *Panicum* community may be occupied by *Zygophyllum* spp. or *Hammada elegans* in the eastern zone of Arabia.

2. *Lasiurus hirsutus* community

Lasiurus hirsutus (Arabic: Dha'ah) is a perennial desert grass which grows in large dense thickets. It grows on alluvial plains and in wadis, attaining a height of 1 m. The plant is not as widesprad as *Panicum*. *Lasiurus* is a common associate in the communities dominated by *Panicum turgidum* and *Pennisetum divisum*. It dominates a community which abounds in areas with alluvial sandy soils. The habitat of this community is subject to water erosion.

Though the dominant plant has medium palatability in the green state, it is eaten when dry by camels and donkeys. However, there are many associates which are of high palatability and contribute to the grazing capacity of this community. These include *Panicum turgidum*, *Aristida mutabilis*, *A. funiculata* and many other ephemerals.

3. *Cymbopogon schoenanthus* community
C. schoenanthus (Arabic: Idhkhir, Skhabar) is a perennial aromatic grass with filiform leaves. The plant grows in tufts and the annual growth takes place from the base of the tuft. This species is extemely drought- and heat-resistant. It is a C_4 plant (Ziegler *et al.* 1981), palatable to sheep and goats in the green state and is also acceptable when dry. Another species, *C. parkeri* grows in the eastern zone of the surveyed territory and dominates a plant community abounding in shallow depressions with fine-textured shallow surface deposits.

C. schoenanthus dominates a community in the ridges of the Hejaz and Oman mountains. It abounds on shallow eroded water courses and stony gullies. The plant cover is low. The palatable associates include *Indigofera spinosa*, *Blepharis ciliaris* and ephemeral grasses. Non-palatable associates comprise *Rhazya stricta*, *Forsskalea tenacissima* and *Euphorbia* spp.

4. *Pennisetum divisum* community
P. divisum (Arabic: Thaymoum) is a soil-binding perennial grass which forms hillocks of considerable size. It is less palatable as is evident from its prosperous growth in habitats where *Panicum turgidum* is overgrazed. However, the range importance of this community is mainly due to the more palatable associates. *P. divisum* dominates a community of wide geographical distribution in the surveyed area. The tree and shrub layer may be represented by *Acacia ehrenbergiana*, *Ziziphus* spp. and *Lycium shawii*. Palatable associates include *Panicum turgidum*, *Rhanterium epapposum*, *Stipagrostis plumosa*, *Eleusine compressa* and *Lasiurus hirsutus*. Among the annuals growing in this community and contributing to its range potential in the rainy season may be mentioned: *Schismus barbatus*, *Erodium laciniatum*, *Launaea* spp., *Centaurea* spp. and *Plantago* spp.

5. *Stipagrostis plumosa* community
Stipagrostis plumosa (Arabic: Nassy) is a perennial desert grass which grows in tufts. The plant produces silvery inflorescences. It accumulates very small sand heaps, not exceeding a few centimetres in height. The plant is a common associate in numerous desert plant communities. It dominates a community which abounds in sandy plains, especially at the feet of dunes in the Nafud, Ad-Dahnaa and the Rub' Al-Khali. It occurs also in wind-eroded habitats where there is a thin sheet of sand cover. The plant cover is thin. Hence, the range potential of this community is limited, especially in the dry season. The common associates include: *Moltkiopsis ciliata*, *Astragalus spinosus*, *Helianthemum lippii*, *Heliotropium bacciferum* and *Cornulaca* spp. After rains, a considerable number of ephemeral species

appear and contribute to the major part of the range capacity of this community.

6. *Cyperus conglomeratus* community
The dominant plant is a densely tufted, pale green perennial herb with coarse rootlets heavily covered with felt-like hairs. It is a multiform species with many varieties in the different countries in the surveyed territory. The plant is a sand dweller in both coastal and inland sandy habitats. The community dominated by *C. conglomeratus* is widespread in the sand formations of the study area. It occurs on aeolian sands with dunal landforms in many regions, e.g. Nafud, Ad-Dahnaa, Rub' Al-Khali, and sand formations in the Arabian Gulf and Gulf of Oman coasts. The dominant plant may form pure stands or have psammophytic associates including *Moltkiopsis ciliata*, *Panicum turgidum*, *Calligonum comosum* and *Monsonia nivea*. After rainy spells, numerous ephemeral plants grow covering the ground surface between the perennials. These ephemerals include: *Cutandia memphitica*, *Neurada procumbens*, *Anthemis deserti*, *Plantago* spp. *Stipa capensis* and *Rumex pictus*.

B. Communities dominated by undershrubs

7. *Rhanterium epapposum* community
R. epapposum (Arabic: Arfaj) is a composite undershrub growing to a height of 1 m. The numerous faggoty branchlets together form a spherical bush. Numerous lateral roots penetrate into the upper 20 to 30 cm of the soil and the tap root descends deeper to reach the permanently wet layer. The plant is favoured by all nomadic animals, sheep, goats and camels. It is eaten largely in the green state.

Rhanterium is the characteristic and most abundant perennial over wide tracts of north central and eastern Arabia. The plant dominates a community inhabiting sandy areas extending from about 30°N to about 25°N. It is never a common plant west of 40°E, but eastwards it extends to within a short distance of the Arabian Gulf coast in Kuwait, Hasa and southern Qatar.

The community occurs in extensive plains with sandy soil, in wide valleys as well as in water runnels. Associated species vary from one locality to another. Among the recorded associates, one may mention *Artemisia inculta*, *Achillea fragrantissima*, *Convolvulus lanatus*, *Cymbopogon schoenanthus* and *Teucrium polium* in the northern sector of Riyadh region, while in the southern sector one finds *Panicum turgidum*, *Lasiurus hirsutus*, *Acacia ehrenbergiana*, *Fagonia indica* and *Lycium shawii*. In the Summan, Kuwait and southern Qatar, the associates include *Stipagrostis plumosa*, *Moltkiopsis ciliata*, *Panicum turgidum*, *Pennisetum divisum* and *Zygophyllum qatarense*.

During spring in rainy years, numerous ephemerals appear including *Medicago aschersoniana*, *Arnebia decumbens*, *Neurada procumbens*, *Horwoodia dicksoniae*, *Schimpera arabica*, *Anthemis deserti*, *Calendula arvensis* and *Stipa capensis*.

8. *Artemisia inculta* community
Artemisia inculta Asso (Arabic: Shih, Bo'aitheran) is a suffrutescent composite plant with fragrant odour. It is favoured by sheep, but is eaten also by goats and camels. *A. inculta* dominates a community of widespread occurrence in northern Arabia, mainly to the north of 24°N latitude. It is of common occurrence in the Hajarah plain and its wadis along the Iraqi borders. The community is subject to overgrazing. The common associates include *Achillea fragrantissima*, *Hammada elegans*, *Anvillea garcini*, *Helianthemum lippii*, *H. ellipticum* and *Astragalus spinosus*. A few trees of *Ziziphus lotus* and *Acacia ehrenbergiana* are found in this community. The common annuals include *Savignya parviflora*, *Koelpinia linearis*, *Schismus barbatus*, *Aizoon hispanicum* and *Aaronsohnia factorovskyi*.

9. *Hammada elegans* community
H. elegans (Bunge) Botsch. (=*Haloxylon salicornicum*) (Arabic: Rimth) is a suffrutescent succulent chenopod with thick jointed branches. The plant has a wide geographical distribution in the Arabian Peninsula. It grows on degraded plains, on the alluvial wadi terraces and on slighlty saline soils indicating its xerohalophytic affinity. It is palatable mainly to camels. The plant is able to form mounds of fine material around its body. The mound usually supports many annual species in the rainy season. The plant has a wide geographic range, being recorded in almost all regions.

In the northern part of the surveyed territory, the community occurs in habitats subject to severe erosion, mainly by wind. The main associates include *Stipagrostis plumosa*, *Lasiurus hirsutus*, *Astragalus spinosus*, *Zilla spinosa*, *Salsola tetragona* and *Traganum nudatum*. It is to be noted that in Qassim, where the area occupied by *H. elegans* is ploughed for cultivation and then left fallow, *Salsola baryosma* replaces *H. elegans* and grows profusely. *S. baryosma* is non-palatable and is usually avoided by sheep grazing in the fallow land.

In the Red Sea coastal plain, the main associates include *Acacia tortilis*, *Panicum turgidum* and *Lycium shawii*. On the other hand, in saline habitats of the Arabian Gulf zone, the main associates are halophytes including *Seidlitzia rosmarinus*, *Suaeda vermiculata* and *Salsola vermiculata*. In the rainy season numerous ephemerals appear including *Stipa capensis*, *Bassia eriophora*, *Rumex vesicarius*, *Savignya parviflora*, *Trigonella stellata*, *Picris radicata*, *Medicago laciniata*, *Anastatica hierochuntica* and *Matthiola longipetala*.

Degradation of this community due to overgrazing which results in severe wind erosion leads to the dominance of *Rhazya stricta* in some localities as in Wadi Al-Batin (29°N, 49°E). The soil supporting this vegetation dominated by *H. elegans* and *Rhazya stricta* is formed of mixed gravel of limestone origin. Wadis which support this community are always focal points for the numerous livestock. The palatable species are overgrazed, sometimes beyond their capacity to regenerate. The common associates include: *Anabasis setifera*, *A. articulata*, *Heliotropium bacciferum*, *Anvillea garcini* and a few individuals of *Acacia ehrenbergiana* and *Ziziphus lotus*. The widespread annual is *Anastatica hierochuntica*.

In the limestone territory of Al-Hajarah region on the frontiers of Iraq, where the *Artemisia* community is deteriorating due to over-grazing, *Hammada elegans* co-dominates. The community co-dominated by *Hammada elegans* and *Artemisia inculta* has a thin plant cover. It occurs on soils made up of rocks and pebbly fine shallow sediment. Both wind and water erosion are fairly pronounced. The associates include: *Achillea fragrantissima*, *Anabasis articulata*, *Anvillea garcini*, *Astragalus spinosus*, *Traganum nudatum* and *Lycium shawii*. Annuals include: *Savignya parviflora*, *Diplotaxis* sp., *Bassia eriophora*, *B. muricata*, *Koelpinia linearis*, *Plantago ovata* and *Trigonella stellata*.

10. *Indigofera spinosa* community
The dominant plant is a silvery frutescent legume plant with glossy needle-like sharp spines. It is common in the western part of the peninsula, mainly in the Hejaz region. *I. spinosa* dominates a community which occurs in rocky foothills, gravelly slopes and terraces, as well as in habitats with coarse-textured soil, covered with rock fragments and boulders, e.g. upstreams of runnels. The water revenue of this habitat is limited. Among the associates one may mention *Aerva javanica*, *Cymbopogon schoenanthus*, *Rhazya stricta*, *Tephrosia* spp. and a few sparse individuals of stunted *Acacia tortilis*. Goats are the main grazing animals since this community is less accessible to other domestic animals.

11. *Scrophularia hypericifolia* community
S. hypericifolia Wydl is a shrubby plant with somewhat spiny twigs. It is very palatable to sheep and goats, which consume the annual growth of leaves and shoots. The community dominated by this plant is found in sandy areas in the northern part of the Arabian Peninsula. It has not been recorded south of 24°N latitude or in the western part of the peninsula. The community occurs on sandy soils with a deep, more or less homogeneous profile of dunal formation. It exhibits pronounced symptoms of overgrazing. The common associates include; *Panicum turgidum*, *Artemisia monosperma*, *Rhanterium epapposum*, *Calligonum comosum* and *Asthenatherum forsskalii*.

C. Communities dominated by shrubs and trees

The presence of trees and shrubs is usually limited to localities of rugged topography receiving run-off water. The common desert trees and shrubs in the surveyed area include: *Acacia* spp., *Ziziphus* spp., *Lycium shawii*, *Leptadenia pyrotechnica*, *Calotropis procera*, *Commiphora* spp. and *Maerua* spp. *Prosopis* spp. mainly occur in the Arabian Gulf zone. *Haloxylon persicum* is common in sandy areas of the Qassim region and northwards, while *Calligonum comosum* grows in sandy areas in the main sand formations in the surveyed region, e.g. in the U.A.E.

The genus *Acacia* comprises numerous species in the surveyed territory.

They grow in various habitats located at various altitudes above sea level. Some of them dominate either desert communities at altitudes less than 1000 m above sea level or mountainous plant communities at altitudes higher than 1000 m. The desert plant communities comprise those dominated by *Acacia tortilis*, *A. ehrenbergiana*, *A. hamulosa*, *A. raddiana* and *A. mellifera*. Mountainous *Acacia* communities are dominated by *A. seyal*, *A. asak* and *A. ethbaica*.

Trees or shrubs of the genus *Acacia* are usually trimmed by camels. Goats too can make use of the accessible parts of the plant. The leaves and fruits may be shaken down with crooks by the Bedouins to feed their animals.

12. *Acacia tortilis* community

The Arabic name of *Acacia tortilis* (Forssk.) Hayne is Samr. This is the most widespread community in the surveyed territory. The community abounds in terraces of wide wadis, shallow and narrow valleys and on gentle slopes. The ground surface is rocky and gravelly. The rock fragments are underlain by compact fine sediment. Due to the local topographical features and compactness of the soil below the blocks and boulders, the water resources in this habitat are limited. This habitat is subjected to considerable wind and water erosion. In the eastern sector of the surveyed territory, e.g. in Qatar, the *Acacia tortilis* community abounds in depressions which have shallow, coarse-textured soil. Stones and gravel cover a considerable part of the ground surface.

The plant cover in this community is low. However, it increases in the rainy season, due to the appearance of ephemerals. In the Hejaz region the common associates include *Lycium shawii*, *Acacia hamulosa*, *A. raddiana*, *A. ehrenbergiana*, *Indigofera spinosa*, *Rhazya stricta* and *Panicum turgidum*. In the eastern zone (Qatar), the common associates comprise *Lycium shawii*, *Panicum turgidum*, *Eleusine compressa*, *Zygophyllum qatarense*, *Convolvulus deserti* and *Rhanterium epaposum*. In Oman, this community is widespread at various altitudes, in wadis, plains, slopes and narrow runnels. The main associates include *Euphorbia larica*, *Rhazya stricta*, *Taverniera glabra*, *Convolvulus* spp., *Zygophyllum* spp., *Lycium shawii* and *Ziziphus spina-christi*.

13. *Acacia ehrenbergiana* community

The Arabic name of *A. ehrenbergiana* Hayne is Salam. The community dominated by this species has a wide geographical amplitude. However, it has a narrower ecological range than that of the community dominated by *A. tortilis*. The community occurs in alluvial plains with sandy soil or wadis and wide runnels with deep deposits, usually of alluvial nature. Associates vary in the different phytogeographical regions of Arabia. The common associates include *Acacia tortilis*, *Lycium shawii*, *Ziziphus* spp., *Aerva javanica*, *Cassia italica*, *Pennisetum divisum*, *Salvia aegyptiaca*, *Heliotropium* spp., *Leptadenia pyrotechnica*, *Francoeuria crispa* and *Rhazya stricta*.

14. *Acacia hamulosa* community
This community has a limited geographical distribution, occurring mainly in the western sector of the surveyed territory. It abounds on slopes and in narrow runnels in the Hejaz mountains. Its habitat is severely affected by erosion and the ground surface is covered with conglomerates and boulders. Grazing, mainly by goats, is practised in this community. This is clear from the overgrazed individuals of *Tephrosia nubica* and the stunted growth of other palatable plants. There are numerous associates in this community. The tree and shrub layer is represented by *A. hamulosa* together with *Ochradenus baccatus*, *Acacia tortilis*, *Lycium shawii*, *Commiphora opobalsamum* and *Acacia mellifera*. The common suffrutescent species include *Indigofera spinosa*, *Pergularia tomentosa* and *Tephrosia apollinea*. Some species adapted to rocky habitats are recorded in this community. These species are *Hibiscus micrantha* var. *genuinus*, *Solanum albicaule*, *Jasonia montana*, *Teucrium leucocladum*, *Valantia hispida*, *Cometes abyssinica*, *Ephedra aphylla* and *Rhus tripartita*.

15. *Acacia raddiana* community
The Arabic name of *A. raddiana* is Talh. This community occurs in wadis and wadi terraces of the Hejaz mountains. It has almost the same geographical distribution as the community dominated by *Acacia hamulosa* but with different habitat features. The habitat supporting the *A. raddiana* community has deep fine sediments. The ground surface is covered with gravel, stones and conglomerates, usually with different colours but not desert-varnished. *A. raddiana* attains a height of 6 to 8 m and its growth forms open forests; the tree and shrub layer is most conspicuous. The main associates include *Acacia tortilis*, *Lycium shawii*, *Indigofera spinosa*, *Cassia italica*, *Morettia canescens* and *Forsskalea tenacissima*. Cutting of *Acacia raddiana* is not uncommon. The habitat of this community is easily accessible to grazing animals and is usually overgrazed.

16. *Acacia mellifera* community
The Arabic name of *A. mellifera* is Kitter. This community is confined to the western sector of the study territory. It occurs at altitudes above 500 m and usually abounds on precipitous cliffs. This community has low cover and the palatable species are overgrazed. The common associates include *Acacia hamulosa*, *Commiphora opobalsamum*, *Tephrosia nubica* (grazed), *T. uniflora* (grazed), *Pimpinella cretica*, *Lindenbergia sinaica*, *Trichodesma ehrenbergii* and *Cleome brachycarpa*.

17. *Prosopis cineraria* community
The Arabic name of *P. cineraria* is Ghaf. The plant is palatable and its fruits present good fodder for animals. The community dominated by this plant is confined to the eastern zone of the studied territory. It occurs on sandy formations in the United Arab Emirates and in wadis crossing the mountains of Oman. The tree and shrub layer is represented by the dominant plant and other weakly represented species such as *Lycium shawii* and

Calligonum comosum. Panicum turgidum and *Indigofera* sp. occur in this community. In the rainy season, there is ample room for the growth of innumerable seedlings of ephemeral species. The accessible branches are usually grazed by camels and goats. The shed fruits are eaten by grazing animals.

18. *Calligonum comosum* community
The Arabic name of *C. comosum* is Arta. *Calligonum comosum* L'Her. is a low multicauline shrub, forming clumps of which the aphyllous branches are in keen demand by camels. The plant has a wide geographical amplitude, but its presence is confined to sandy soils, particularly on the top of dunes. The plant dominates a community with low plant cover in the sandy habitats of the Rub' Al-khali, Nafud, Ad-Dahnaá and in many localities in the United Arab Emirates. The dominant plant exhibits local abundance in localities with loose, coarse sandy soils. The main associates include *Panicum turgidum*, *Cyperus conglomeratus*, *Asthenatherum forsskallii* and *Artemisia monosperma*. Many ephemeral species appear in the rainy season, contributing to the range potentiality of this community.

II. HALOPHYTIC PLANT COMMUNITIES

Salt-affected lands occupy vast areas in the study territory. In addition to the coastal and inland salines, there are considerable areas with salinized soil due to anthropogenic activities. These salines harbour halophytic plants. The main halophytic plant communities with considerable range potentiality will be enumerated in the present study.

19. *Avicennia marina* community
The Arabic names of *A. marina* are Shourah and Girm. The dominant plant attains considerable heights, especially in southern latitudes. It forms pure populations and fringes the coast along the Red Sea shore and the Arabian Gulf. On the latter, it is confined to particular sites. *A. marina* is usually browsed by camels, especially during low tide.

20. *Aeluropus lagopoides* community
A. lagopoides (Arabic: Ikrish) is a perennial grass with widespread distribution in the coastal marshes of the surveyed territory. The community occurs on saline flats with fine-textured soil. Though the dominant plant occurs in pure stands in many localities, it may occasionally be mixed with *Limonium axillare*. *Aeluropus* is usually grazed by camels.

21. *Suaeda* spp. communities
Some *Suaeda* species dominate plant communities in the study area including *S. monoica*, *S. vermiculata*, *S. fruticosa* and *S. pruinosa*. These communities abound in coastal and inland salines. *Suaeda vermiculata* is a palatable plant grazed by camels. Its Arabic name is *Suweid*. This plant

forms sizeable mounds of fine-textured soil. This community occurs in the eastern zone of the study area. *Suaeda monoica* dominates a community occurring in inland salines in the western zone of the study territory, while *S. pruinosa* dominates a plant community in central and northern Arabia. The dominant species as well as some associates of these communities are usually browsed by camels.

22. *Salsola* spp. communities
Several *Salsola* species are recorded in the study area, including *S. tetrandra*, *S. tetragona*, *S. vermiculata*, *S. baryosma* and *S. cyclophylla*. Some of these species dominate plant communities occurring usually in inland salines and on degraded soils.

23. *Seidlitzia rosmarinus* community
The community dominated by this plant is of wide geographical distribution. It abounds mainly in the lower parts of valleys which pour their water into the Red Sea, or in wadis where soil salinity is high in the eastern zone and in central Arabia. The common associates include *Tamarix* sp., *Suaeda* sp., *Aeluropus lagopoides* and *Lagonychium farctum*.

24. Other halophytic plant communities
There is a number of plant communities dominated by halophytic species which inhabit inland or coastal salines. However, they are of low range potentiality. These include communities dominated by: *Halocnemum strobilaceum*, *Arthrocnemum glaucum*, *Limonium axillare*, *Halopeplis perfoliata*, *Halopyrum mucronatum*, *Sporobolus spicatus*, *S. arabicus* and *Tamarix* spp.

III. MOUNTAINOUS PLANT COMMUNITIES

The environmental conditions in the western highlands and Oman mountains are widely different from those in the rest of the surveyed territory. This is mainly due to high altitude, with its consequences on air temperature, and to relatively high rainfall without a prolonged dry period. Such environmental conditions are reflected on the plant life, which is relatively rich in species and plant cover. It is regretted that there is not much data or information on the vegetation of these mountains. However, the most common communities, with considerable contribution to range potentiality in the study area, will be stated briefly.

25. *Juniperus excelsa* community
The Arabic name of *J. excelsa* is Ar'ar. The community dominated by *J. excelsa* is widespread in the Asir mountains. In Jabal Al-Akhdar, there is another species, viz. *J. macropoda*. It occurs on the slopes and mountains above 1700 m altitude. The density and vigour of the tree vary in the different localities, depending on human activities. Among the associates of

this community, there are important range grasses, including *Themeda triandra* and *Hyparrhenia hirta*. *T. triandra* Forssk. is a densely tufted grass with upright stalks growing annually from buds at the base of the tuft. During the growth phase, it is appreciated by sheep, goats and bovines. *Hyparrhenia hirta* (L.) Stapf is a tufted grass, 50–100 cm tall, with branched culms. It grows in the loamy interstices of rocks and large boulders. Although the plant is liked by sheep, the non-accessibility of its habitat in many localities protects it from being overgrazed. Other associates include *Euryops arabicus*, *Lavandula dentata*, *Dodonea viscosa*, *Mentha longifolia*, *Euphorbia schimperi* and *Conyza incana*. Rock crevices in shaded places support numerous ferns. The *Juniperus* trees are usually festooned with lichens. There is another species, viz. *J. phoenicea*.

26. *Olea europaea ss- africana* community
The Arabic name of this *Olea* sp. is A'tam. It is an evergreen tree, heavily browsed by animals and cut by natives, who use its branches as a substitute for the toothbrush obtained from *Salvadora persica*. The community dominated by this plant occurs at levels above 1500 m altitude and is usually found along the transmontane valleys. Among the associates of this community may be mentioned *Trachonanthus camphoratus*, *Pistacia palaestina*, *Dodonea viscosa*, *Euclea kellav* and *Carissa edulis*. On the Jabal Al-Akhdar, there is a community of *Reptonia mascatensis–Olea africana* at elevations of 1350–2300 m. Among the associates there are *Sageretia spiciflora*, *Ziziphus* sp., *Juniperus macropoda*, *Ebenus stellatus*, *Acacia eburnea*, *Dodonea viscosa* and *Ephedra intermedia*.

27. *Acacia asak* community
The Arabic name of *A. asak* is A'sak. The community occurs on the hill-sides and plateaus between 1000 and 1500 m altitude. *Acacia ethbaica*, *Commiphora* spp., *Grewia velutina* and *Ficus salicifolia* are common associates. *Blepharis persica*, a palatable plant, is very common. Numerous grasses associated with this community include *Andropogon distachyos*, *Aristida coerulescens*, *Stipagrostis obtusa*, *S. hirtigluma*, *Bromus fasciculatus*, *Cenchrus ciliaris*, *Chrysopogon aucheri*, *Cymgopogon schoenanthus*, *Danthoniopsis barbata*, *Eragrostis papposa*, *Hyparrhenia hirta*, *Pennisetum setaceum* and *Tetrapogon villosus*.

28. *Ficus salicifolia* community
The community dominated by this species occurs on the slopes and along the transmontane valleys. The common associates include *Hyparrhenia hirta*, *Aerva lanata* and *Capparis cartilaginea*. Ferns are found in crevices between the rocks.

29. *Acacia seyal* community
This community is found at levels above 1800 m altitude. It occurs on hills and plains with sandy soil derived from the weathering of the granites.

The main associates are *Kleinia odora* (which sometimes co-dominates the community), *Ficus salicifolia*, *Chrysopogon aucheri*, *Panicum turgidum*, *Pennisetum divisum*, *Acacia asak*, *Lavandula* sp. and *Cucumis figarei*.

CONCLUDING REMARKS

The distribution of the plant communities recognized in the surveyed territory is controlled by two main factors, viz. the geographical location and physiographic features. These two factors influence the distribution, establishment, growth and regeneration of the plants through their effect on the various environmental factors. Rangeland management should depend on a thorough knowledge of the interaction between landscape (local topography), soil, vegetation, climate and grazing animals, which affect the productivity of the rangelands.

It is a well known fact that variability in rainfall increases as mean annual rainfall decreases, so that the driest climates have the least predictability. Under such conditions, which prevail in the surveyed territory, the redistribution of water by runoff is of vital importance to plant life. Also, the physical attributes of the soil are influenced by topography. The plant cover in the desert is closely related to the physical properties of the soil and the water resources.

Management of rangelands should take into consideration the changes manifested by the desert vegetation. These changes include: (a) rhythmic changes, (b) irregular changes, (c) developmental changes and (d) retrogressive, man-made changes.

The effects of man on the desert vegetation may be direct, on the vegetative cover itself, or indirect, through their influence on the other components of the ecosystem. The direct effects include the traditional practices of cutting or uprooting ligneous species for firewood, overgrazing, exploitation of raw material and eradication of plants as a result of construction activities, especially that of roads. The indirect effects are more drastic. Human activities, through their effect on the physical components of the fragile desert ecosystem, contribute to the disruption of the natural equilibrium among the components of the ecosystem. Usually, the main component subject to degradation is the soil, which is the basic resource of the ecosystem. This results inevitably in retrogressive changes in the vegetation.

It is to be noted that human activities accompany the development process in the surveyed territory. The human influence on the desert landscape has increased recently, due to the rapid development. The increasing population in these countries with expanding consumption puts significant pressures on the desert. Large areas have been mechanically destroyed in connection with the rapid development. There is a great development in the technology of oil and natural gas transport through pipelines and construction of modern highways to replace the very poor trails. Mining activities are continuously increasing and quarrying is widespread in the desert. These processes, in addition to their direct effects causing eradication of considerable acreage of plant growth, engender changes in the vegetation due to their infuence on the components of the ecosystem. In many cases, the soil is severely eroded, runoff increases and less water is allowed to permeate to deep soil layers. Hence, retrogressive changes are apt to occur.

The continuous increase of water points along roads and pipelines facilitate overgrazing. The life of the Bedouins has been changing because of their greater mobility. Good-quality water is now trucked from village wells to camps far out in the desert. People, animals and food are now transported over the desert terrains far more rapidly than by camel and donkey transportation used in former years. This permits movement when forage is exhausted. Such a practice is a major cause for the ever increasing area of overgrazed and disturbed range. The number of cars and trucks used by bedouins in the desert is increasing, causing the removal of plant cover and soil erosion. From the air, one can notice that the desert is heavily dissected with the tracks of cars. The degradation due to cars remains for a long time and is usually irreversible. The exodus of city inhabitants to the desert on weekends is a common tradition in the countries of the surveyed territory. This results in the direct destruction of the plant cover as well as the degradation of the ecosystem by accelerating soil erosion. In general, these processes cause the establishment of desert pavements with a gravel-strewn surface in some parts, as well as the accumulation of sand embankments and sand dunes in others. Desert pavements are devoid of plant cover and dunes may remain mobile and barren or are covered by a thin plant cover.

Rangeland management in the countries of the surveyed territory is not an easy task, on both local and regional levels. This is attributed to numerous factors, including physical and social factors as well as the lack of proper ecological information about the rangelands, on both the species and community levels. The severe physical environmental conditions are represented by low, unreliable rainfall and remarkably high evaporative power of the atmosphere. Degraded soils aggravate the problem.

During the last few decades, these countries have witnessed rapid socio-economic changes. These changes, in different ways, have exerted considerable pressure on the natural resources, especially water resources (*cf.* Batanouny 1978). For a range manager, there is a great need for continuous evaluation and replanning when confronted with new situations and changing patterns of resources and social changes.

The lack of data is among the main problems confronting range management in the surveyed territory. Even the available data are given in reports submitted by various consultants and these lack a consistent set of standard definitions. This lack inevitably leads to considerable difficulty in rationalizing and comparing the results of one survey with another. The data collected must be unified in a manner suitable for ready retrieval of specific types of information.

Studies on the rangelands and their development in the surveyed territory have been carried out mainly by foreign experts and there has been a lack of local participation in the design and evaluation of the development of range resources. In addition to the language barrier, there are many socio-economic aspects, which need to be investigated. It will be fruitful only if the experts work in close cooperation with local professionals.

There is a great lack of relevant data and statistical information from the past. A knowledge of the history of rangeland use and management (e.g. the *hema* system) in the surveyed territory is of help in the assessment of many factors involved in the development of range resources.

LITERATURE CITED

Annual Statistical Abstract of Qatar. 1984. Central Statistical Organization, Doha, No. 4, July 1984.

Anonymous. 1973. *The Middle East and North Africa.* Europa Publ. Ltd., London.

AOAD (Arab Organization for Agricultural Development). 1981, 1983. *Yearbook of agricultural statistics,* vols 1 and 3, Khartoum.

Batanouny, K. H. 1975. Certain factors involved in the evaluation of lands in arid and semi-arid zones. p. 79–96. Proc. Int. Seminar on the Evauation of Lands in Arid and Semiarid Zones of Latin America. IILA, Rome, 4–7 February 1974.

Batanouny, K. H. 1978. Socio-economic changes and development of water resources in Saudi Arabia. p. 935–950. *In:* G. C. Vansteenkiste (ed), *Modeling, identification and control in environmental systems.* North-Holland Publ. Co., Amsterdam.

Batanouny, K. H. 1979. Vegetation along the Jeddah-Mecca road: Pattern and process as affected by human impact. *J. Arid Environ.* **2:** 21–30.

Batanouny, K. H. 1981. *Ecology and flora of Qatar.* Scientific and Applied Research Centre, Univ. of Qatar.

Batanouny, K. H. and N. A. Bashin. 1983. Plant communities along the Medina-Badr road across the Hejaz mountains, Saudi Arabia. *Vegetatio* **53:** 33–43.

Batanouny, K. H. and N. A. Baeshin. 1983. Plant communities along the Medina-Badr road across the Hejaz mountains, Saudi Arabia. *Vegetation* **53:** 33–43.

Carbon, B. A. 1975. Redistribution of water following precipitation on previously dry sandy soils. *Austral. J. Soil Res.* **13:** 13–19.

El-Khatib, A. B. 1974. Seven green spikes, water and agricultural development. Ministry of Agric. and Water, Riyadh.

Kernick, M. D. 1966. Report to the Government of Kuwait on plant resources, range ecology and fodder plants introduction. Rep. No. 2181, KUW/TE/PL, PL/TA/11, FAO, Rome.

FAO. 1973. Agricultural development in the United Arab Emirates. Report on present position, prospects and priorities. Prepared by FAO Techn. Adv. Mission, Near East Regional Office, Cairo.

FAO. 1974. Near East. Statistical Directory. Near East Regional Office, Cairo.

FAO. 1979. Pastoral systems in the ECWA Region: Some policy considerations. Joint ECWA/FAO Agriculture Division, Mimeo, Rome.

Giacomini, G. and G. De Marco. 1975. La cartografia integrata per le aree semi desertiche unesperienz in Arabia Saudita. p. 329–397. Proc. Int. Seminar on the Evaluation of Lands in Arid and Semi-arid Zones of Latin America. IILA Rome, 4–7 February 1974.

Halwagy, R. and M. Halwagy. 1974. Ecological studies in the desert of Kuwait. II. The vegetation. *J. Univ. Kuwait (Sci.)* **1:** 87–95.

Heady, H. F. 1963. Report to the Government of Saudi Arabia on grazing resources and problems. ETAP Report No. 1614 (Project SAU/TE/PL).

ITALCONSULT. 1969a. Water and agricultural development surveys, Areas II and III. Final Report, Specific Report on grazing resources inventory, submitted to the Ministry of Agric. and Water, Riyadh.

ITALCONSULT. 1969b. Water and agricultural development surveys, Area IV. Final Report, Specific Report on grazing resources inventory, submitted to the Ministry of Agric. and Water, Riyadh.

Johns, G. G., D. J. Tongway and G. Pickup. 1984. Land and water processes. p. 25–40. *In:* G. N. Harnington, A. D. Wilson and M. D. Young (eds), *Management of Australia's rangelands.* Commonwealth Scientific and Industrial Research Organization (CSIRO), Australia.

Kassas, M. 1966. Plant life in deserts. p. 145–180. *In:* E. S. Hills (ed), *Arid Lands.* Methuen, London; UNESCO, Paris.

Kernick, M. D. 1966. Report to the Government of Kuwait on plant resources, range ecology and fodder plants introduction. Rep. No. 2181, KUW/TE/PL/11. FAO, Rome.

Mandaville, J. P., Jr. 1977. Plants. p. 229–265. *In: The scientific results of the Oman Flora and Fauna Survey 1975.* Journal of Oman Studies, Special Report. Ministry of Information and Culture, Mascat, Oman.

Meigs, P. 1953. World distribution of arid and semi-arid homoclimates. *In: Reviews of research on arid zone hydrology* **1:** 203–210. UNESCO, Paris.

Monod, Th. 1954. Mode contracté et diffus de la végétation Saharienne. p. 35–44. *In:* J. L. Cloudsley-Thompson (ed), *Biology of Deserts.* Institute of Biology, London.

Parson, B. 1968. Agricultural and water resources. The Great Nafud sedimentary basin, Kingdom of Saudi Arabia, vol. 2. The agricultural resources. Report submitted to the Ministry of Agric. and Water, Riyadh.

Popov, G. and W. Zeller. 1963. Ecological survey: Report on the 1962 survey in the Arabian Peninsula. U.N. Special Fund, Desert Locust Project, Progress Report UNSF/DL/ES6. FAO, Rome.

SOGREAH. 1968. Water and agricultural development studies. Area V. Final report, vol. 5. Agricultural, grazing and fishing. Report submitted to the Ministry of Agric. and Water, Riyadh.

SOGREAH. 1970. Water and agricultural development studies. Area VI. Final report, vol. 4. Agricultural and grazing potential. Report submitted to the Ministry of Agric. and Water, Riyadh.

Statistical Yearbook of Oman. 1978 and 1980. Development Council, Technical Secretariat, Directorate General of National Statistics. Sultanate of Oman, Nos. 7 and 9.

Vesey-Fitzgerald, D. F. 1955. Vegetation of the Red Sea coast south of Jeddah, Saudi Arabia. *J. Ecol.* **43:** 477–489.

Vesey-Fitzgerald, D. F. 1957a. The vegetation of the Red Sea coast north of Jeddah, Saudi Arabia. *J. Ecol.* **45:** 547–562.

Vesey-Fitzgerald, D. F. 1957b. The vegetation of central and eastern Arabia. *J. Ecol.* **45:** 779–793.

Walter, H. 1955. Die Klimadiagramme als Mittel zur Beurteilung der klimaverhaeltnisse fuer oekologische, vegetationskundliche und landwirtschaftliche Zwecke. *Ber. deutsch. Bot. Gesell.* **68:** 331–344.

Walter, H. 1963. The water supply of desert plants. p. 199–205. *In:* A. J. Rutter and F. H. Whitehead (eds), *The water relationships of plants.* Brit. Ecol. Soc., London.

Ziegler, H., K. H. Batanouny, N. Sankhla, O. P. Vyas and W. Stichler. 1981. The photosynthetic pathway types of some desert plants from India, Saudi Arabia, Egypt and Iraq. *Oecologia* **48:** 93–99.

Zohary, M. 1973. *Geobotanical foundation of the Middle East.* G. Fischer, Stuttgart.

Ecological monitoring of vegetation in Kuwait's rangelands

SAMIRA A. OMAR, FAISAL K. TAHA AND AHMAD NASSEF

Aridland Agriculture Department, Food Resources Division, Kuwait Institute for Scientific Research, P.O. Box 24885, Safat, Kuwait

ABSTRACT

Rangelands in Kuwait provide renewable goods and services such as forage, fiber, meat, water and recreational pursuits. The resources have a deep sentimental value for Kuwaitis who consider them as an integral part of their natural heritage. On the other hand, rangelands have been subjected to pressures that either altered or destroyed them through overgrazing, uprooting of plants and off-route use by vehicles. Some remedial measures are needed if the resources are to continue to produce.

Research on the current status of vegetation in Kuwait's rangelands has been rather scarce and fragmentary. A more serious effort has been initiated by Kuwait Institute for Scientific Research (KISR) in 1979, when an extensive vegetation survey has been conducted at the *Rhanterium*/*Cyperus* steppe. Ecological monitoring of vegetation in this steppe included assessment of 12 protected sites which differed in their characteristics (soil and vegetation). Data collection and monitoring on site were conducted in 1980, 1981 and 1982. Parameters investigated included percent cover, frequency, density, phenology and production. It was observed that sites that enjoyed good protection demonstrated significant increases in plant cover, frequency and density. The average herbage production varied among sites and years. In 1980, 1981 and 1982 the average herbage production was 205.4, 321.1 and 93.6 kg/ha, respectively (dry matter). Recording of the chronological dates for the various developmental stages of plant growth for most range plants, exhibited great variations in growth pattern. The phenological variations were attributed primarily to differences in soil characteristics which either delayed or enhanced the initiation of plant growth.

Extensive field trips were undertaken throughout Kuwait's rangelands to collect, identify and mount native plants. KISR herbarium was established and baseline information and a pictorial book on the flora of Kuwait were prepared.

INTRODUCTION

The climate in Kuwait is very hot and dry during the summer and cool to mild, with rain averaging 112 mm in winter (Halwagy 1986). Data for the Kuwait International Airport (1957–1983) indicate a mean temperature for July and August of 37°C and for January of 14°C. The absolute maximum temperature recorded is in July and August (50.7°C), and the absolute minimum temperature is registered in January (−4°C). Soils of Kuwait

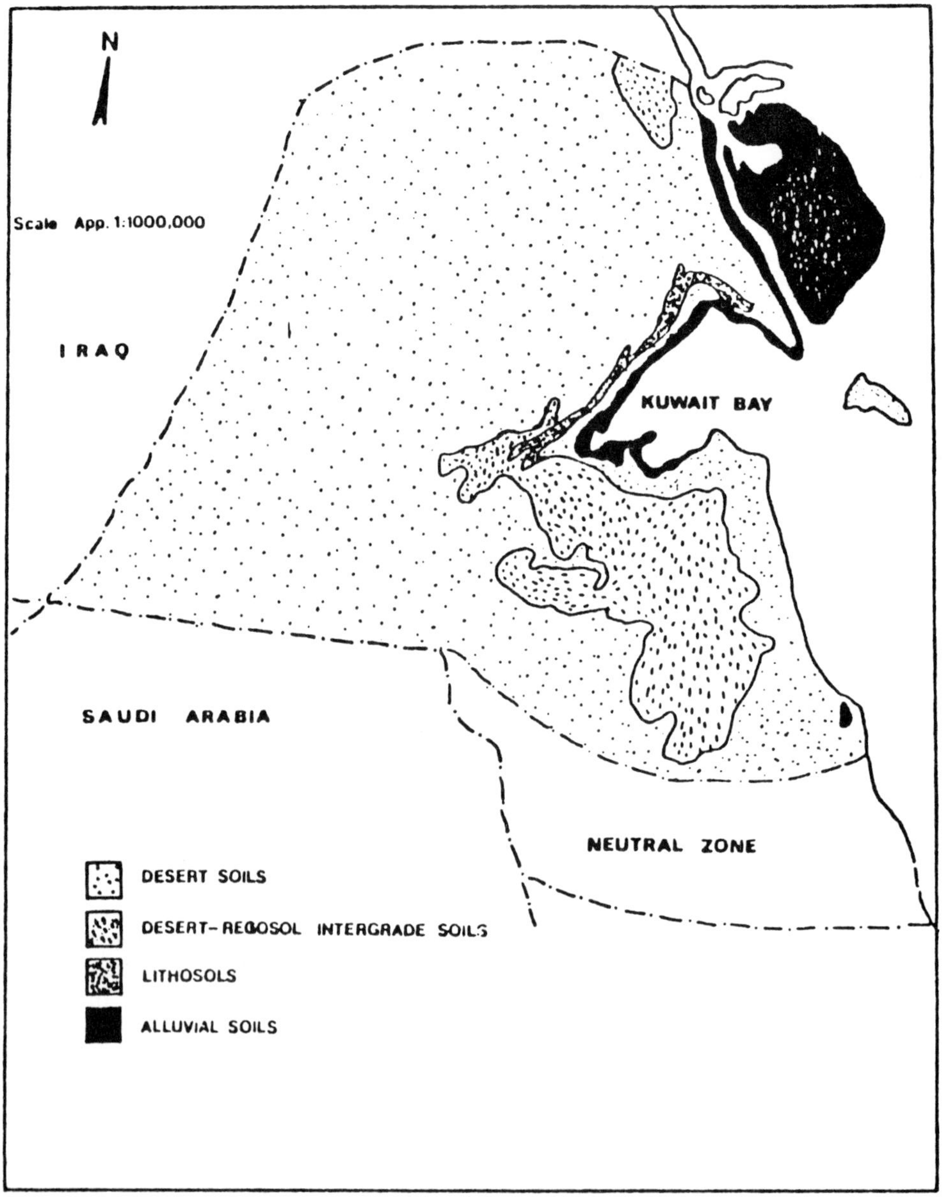

Fig. 1. Major soil groups in Kuwait (Ergun 1969).

were surveyed by Ergun (1969) who recognized four great soil groups (Fig. 1).

1. Desert soils, covering approximately 78% of the country; 2. Desert regosol intergrade soils, (14%); 3. Lithosols, (7%); and 4. Alluvial soils, (1%).

Research on Kuwait's rangelands has been very limited and fragmented in scope. The most significant study on vegetation is that of Halwagy and Halwagy (1974). Their study introduced and distinguished four types of ecosystems: (a) The sand dune ecosystem; (b) The salt marsh and saline depressions ecosystem; (c) The desert plain ecosystem; and (d) The desert plateau ecosystem.

Five major plant communities were recognized (Fig. 2). *Haloxylon salicornicum*, *Rhanterium epapposum*, *Cyperus conglomeratus*, *Zygo-*

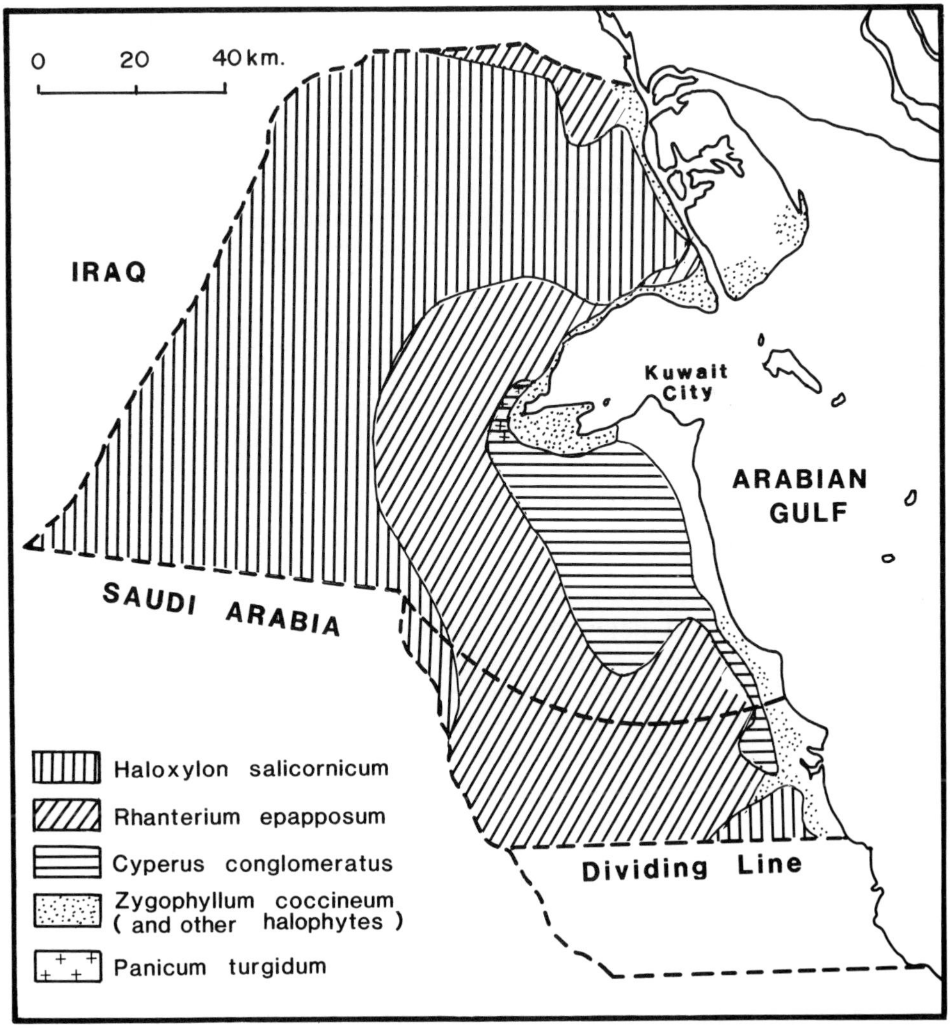

Fig. 2. Vegetative map of Kuwait (Halwagy and Halwagy 1974).

phyllum coccineum, and *Panicum turgidum* communities cover approximately 75% of rangelands in Kuwait. They provide renewable goods and services such as forage, fiber, meat, water and recreational pursuits. The resources have a deep sentimental value for Kuwaitis who consider them as an integral part of their natural heritage. On the other hand, rangelands have been subjected to pressures that either altered or destroyed them, for example, overgrazing, uprooting of plants and off-route vehicles use. Some remedial measures are needed if the resources are to continue to produce.

Research on the current status of the natural vegetation on Kuwait's rangeland has been limited. Data are usually obtained through quantitative ecological studies that measure certain vegetation variables (Dyksterhuis 1949, 1958). These variables show the successional patterns of vegetation production, range condition, and potential and natural herbage structure. Additionally, they reflect the extensiveness of resource degradation and the individual status of current range ecosystems. Thus it becomes possible to specify some remedial action in a suitable range restoration program. Also, the investigation provides information of use as a basis for national policies and programs concerning management and development of rangeland resources.

For the purpose of this study and because there is no comprehensive information on the management and production of grazing land resources in Kuwait, a basic and fundamental approach to resource evaluation was used. Emphasis was placed on the two major grazing areas in Kuwait, the *Rhanterium* and *Cyperus* steppes which cover more than half of Kuwait's rangelands. Quantitative parameters selected for the study are: ground cover, species frequency and density, phenology, and herbage production. Qualitative factors are utilized in the evaluation of sand collection as well as in plant distribution and morphological description.

MATERIAL AND METHODS

Selection of study sites

The reconnaissance survey of Kuwait's major grazing regions (*Rhanterium* and *Cyperus* steppes) delineated 12 major protected sites: Rawdatain, Mutla, Doha, Um-Ghara, KISR Sulaibiya Station, Sulaibiya Transmission Station, Abdaliyah, Minagish, Maqwa, Hoban, A. Al-Jabir Base and A. Al-Salim Base (Fig. 3). The KISR Sulaibiya Station was subdivided into 15 study locations for measurement of vegetation (Fig. 4). The sites varied greatly in soil type, topography, drainage, slope and vegetation cover. Rawdatain is generally sandy loamy in texture, strongly calcareous, with a mixed vegetation of *Haloxylon* and *Rhanterium*. Doha is strongly calcareous, shelly and covered primarily with salt marsh vegetation. KISR Sulaibiya Station is primarily sandy to loamy in texture, undulating to level, and covered with *Rhanterium* intermixed with *Cyperus*. The Sulaibiya Transmission Station is very similar to the KISR site, except that it is fairly flat. The Abdaliyah area varies between sandy loamy to sandy clay loam, with *Rhanterium* as the major vegetation cover. The Minagish study site is

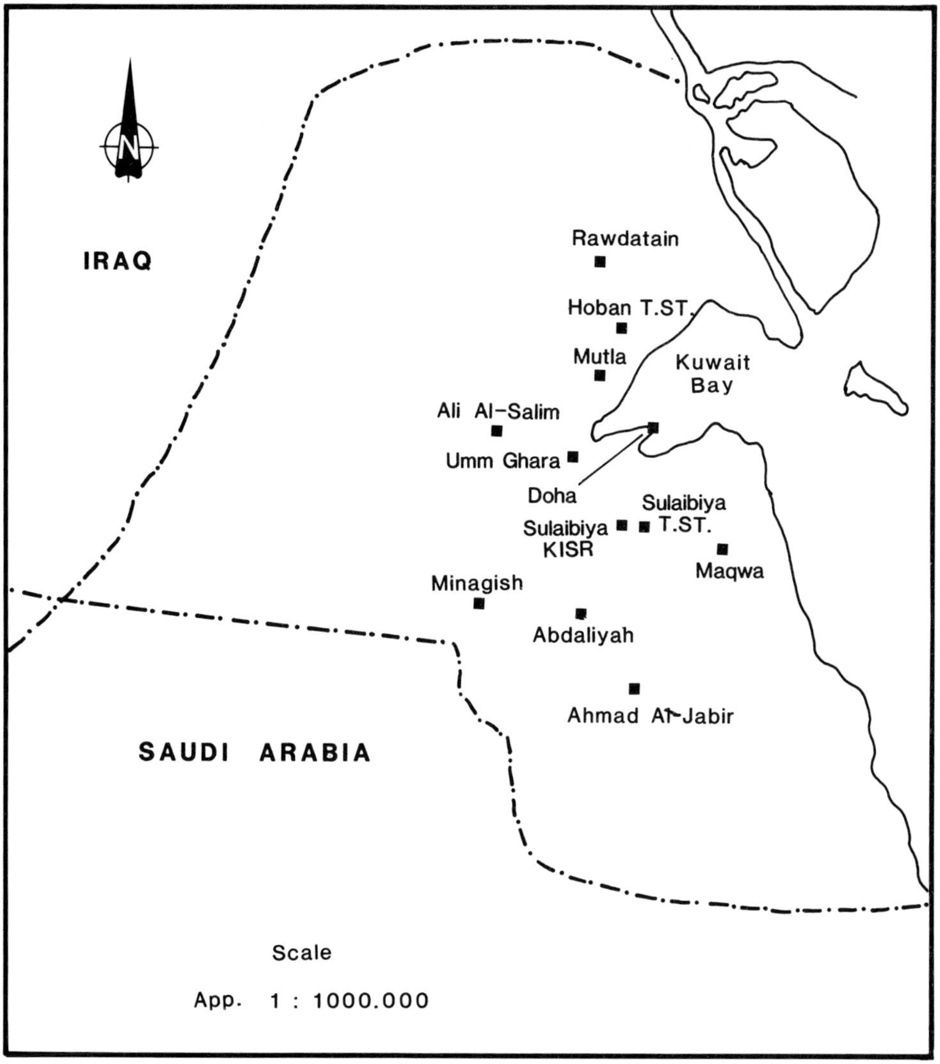

Fig. 3. Location of protected sites selected for vegetation survey studies.

primarily sandy in texture, loose, moderately calcareous with a community of *Rhanterium* and *Cyperus*. The Hoban Transmission Station is similar to the Mutla site with *Rhanterium*. A. Al-Jabir Base is sandy loamy in texture with a mixed vegetation dominated by *Rhanterium* and *Stipagrostis*. A. Al Salim Base is sandy loamy in texture, moderately calcareous with a vegetation of *Rhanterium* and *Stipagrostis*.

Quantitative evaluation of vegetation

Percent Cover. To determine the percent ground cover composition, quadrat-placement method was used (Parker and Pickford 1937; Carnfield 1941; Brown 1954). Ten quadrats, 60 × 30 cm, were randomly placed along

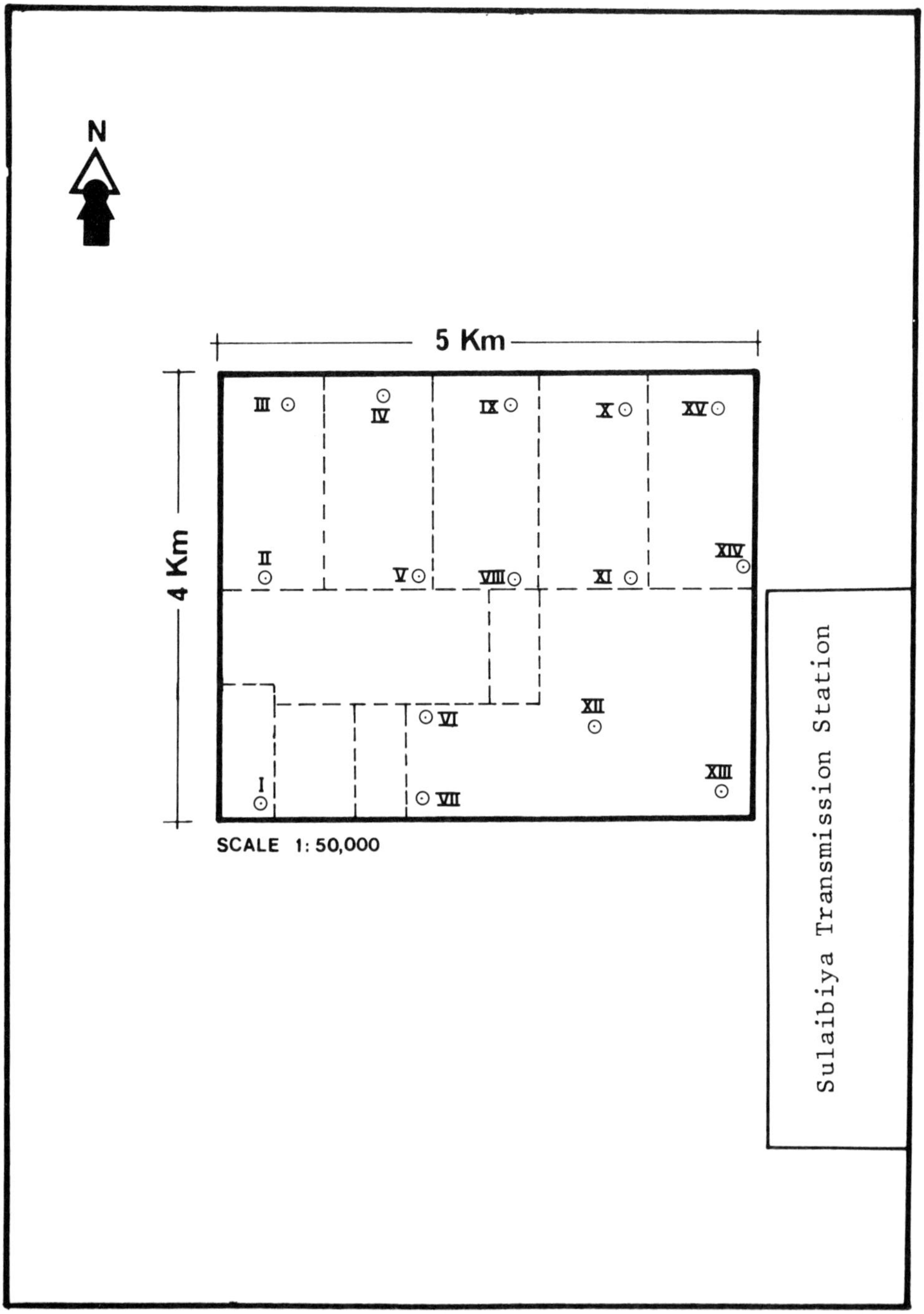

Fig. 4. Site numbers and locations at KISR Sulaibiya Station.

five transect lines (30 m long) repeated four to 15 times depending on the size of the site. Data were recorded for individual vegetation species and material associated with them (bare, litter or gravel).

Frequency. For determination of frequency, data were taken from the same quadrats used for determining percent cover. The presence and absence of species was recorded and percent frequency was computed.

Density. It was determined by counting the individual perennial species within the quadrats (60 × 30 cm) randomly located along five transect lines. Density data were limited to perennial species because of time and manpower limitations.

Production. Vegetation production was divided into three classes: grasses, forbs and shrubs. Grass and forb production was determined by clipping 15 quadrats (30 × 60 cm), randomly located within each study area. The clipping was conducted at or near the peak biomass for these two herbaceous strata, estimated to occur 1 May through 15 May. Grass and forb species were clipped at ground level. Species of individual vegetation stratum were combined and bagged. Clippings were oven dried at 70°C for 24 hours and weighed; production was computed in kg/ha.

Shrub production clipping was initiated in early June to allow for maturation of *Rhanterium*. The clipping was conducted in 1 × 2 m quadrats. Only the current year's growth was clipped and bagged, oven dried at 70°C for 72 hours and weighed.

Phenology. A phenology inventory of important and abundant species was taken at Sulaibiya KISR Station during the study period (1980–1982). It was subdivided into 15 study locations based on soil and topographical variations (Fig. 4). In each substation, the primary plant species were identified and marked with metal labels with specific code numbers. Phenological development was recorded for individual species at weekly intervals throughout the seaon (primarily October through May), according to Bennacchio (1970), and Fisser *et al.* (1976, 1977).

Qualitative evaluation of vegetation

Plant collection and establishment of KISR Herbarium. Extensive field trips were conducted throughout Kuwait's rangelands for plant collection. Plant species were collected from the five prime ecological vegetational zones (Fig. 2). Individual plant species were dried, mounted, identified and placed at the KISR Herbrium. The identification data include family, genera, species, location, date of collection and habitat.

Baseline Information. Data gathered from field observations were pooled to develop baseline information about Kuwait's range plants. The information includes description of habitats, physical attributes, abundance, growth form, life form and uses. A computer program was prepared in order to store the information collected from this survey.

RESULTS AND DISCUSSION

Parameters selected for monitoring and evaluating Kuwait's rangelands emphasized inventory and plant ecological characterization. Data were gathered since 1980 and up to 1985. The quantitative information gathered for all years is considered baseline data for vegetation trend and condition, and included percent cover, frequency, density, production and phenology. Qualitative data were mainly a collection of plants for the herbarium and the preparation of general baseline information.

Due to the large amount of data collected in the survey, and in order to simplify data interpretation, only data for the years 1980, 1981 and 1982 will be discussed in this paper. Results of vegetation monitoring in subsequent years (1983, 1984 and 1985) are stored in the KISR computer.

Cover, frequency and density

Mean percent ground cover of the more important species in the entire KISR Sulaibiya Station from data obtained in 1980 and 1981 at the 15 study

Table 1. Quantitative evaluation of vegetation measurement at Sulaibiya KISR Station.

Parameter	*Cover (%)*		*Frequency (%)*		*Density (%)*	
	1980	*1981*	*1980*	*1981*	*1980*	*1981*
Bare	11.6	30.6				
Litter	6.9	5.5				
Gravel	38.1	4.9				
Total vegetation	43.2	59.0				
Shrubs						
Rhanterium epapposum	5.8	5.3	25.2	21.3	1.4	1.3
Grasses and grasslike						
Asphodelus tenuifolius	0.3	0.4	9.3	5.3		
Cutandia memphitica	6.7	0.8	68.3	12.0		
Cyperus conglomeratus	1.2	2.4	13.3	17.3	0.6	1.2
Iris sisyrinchium	0.0	0.1	0.0	1.3		
Schismus barbatus	0.8	0.9	34.7	12.0		
Stipagrostis plumosa	1.5	0.9	21.6	13.4	4.6	3.8
Forbs						
Arnebia decumbens	0.0	0.1	0.0	2.7		
Astragalus schimperi	0.4	1.9	9.3	33.3		
Atractylis flava	0.5	1.7	20.0	26.7		
Fagonia glutinosa	0.0	0.3	0.0	2.7		
Hymenocarpus nummularis	—	0.2	—	1.3		
Ifloga spicata	0.7	0.9	20.3	28.0		
Koelpinia linearis	1.1	5.8	19.3	53.3		
Launaea mucronata	0.0	0.8	0.3	13.4		
Lotus pusillus	1.5	3.6	38.0	54.7		
Medicago laciniata	0.1	14.6	3.2	78.7		
Moltkiopsis ciliata	0.3	0.1	10.3	1.3		
Plantago boissieri	14.8	16.6	80.5	96.0		
Picris babylonica	0.0	1.1	0.0	16.0		
Polycarpaea repens	0.1	0.4	2.5	8.0		
Silene villosa	0.1	0.1	1.7	1.3		

locations is shown in Table 1. In 1980, gravel was about 38%, bare 12%, litter 7% and vegetation 43% of the ground cover. In 1981, gravel was about 5%, bare 31%, litter 6% and vegetation increased to 59% of the ground cover. In both years, however, *Rhanterium epapposum*, the ecologically dominant species, constitutes 5% of the total vegetation cover, whereas, dominant perennial sedges and grasses, *Cyperus conglomeratus* and *Stipagrostis plumosa* occupied 2 and 1%, respectively, of the ground cover. Table 1 clearly shows that the annual vegetation stratum made up the bulk of the vegetation composition in both years. It occupied more than 50% of the sites surface area. This is considered a fairly good cover for aridlands and possibly places the site in a good range condition.

Frequency data shown in Table 1, differed among and within species.

Table 2 summarizes percent cover for total vegetation inside and outside the fences in 12 study locations. The vegetation pecent cover inside the fences in 1980 varied among locations and ranged between 4.73 and 54.24%, whereas in 1981, it ranged between 17.80 and 59%. The variation among locations indicates the effect of site factors on the percent cover parameter.

Data for 1981 included four additional sites that were well protected, namely; Ali Al-Salim Base, Ahmad Al-Jabir Base, and Hoban and Sulaibiya Transmission Stations. The effect of protection on vegetation is clearly demonstrated in these sites. Data for percent cover of total vegetation showed significant difference between inside and outside of the fences. The cover values for these exclosures as well as for others are considered promising when we realize the harsh environment under which these plants live. The partial protection offered by these exclosures has proved to be very effective in terms of increased vegetation cover. These data also show that, despite the known fact that Kuwait's rangelands

Table 2. Comparison of vegetation cover among sites for 1980 and 1981.

	Total vegetation			
	1980		*1981*	
Site	*Inside*	*Outside*	*Inside*	*Outside*
Doha	4.73a*	4.66a	17.80a	29.90b
Um-Ghara	34.90a	12.01b	—	—
Abdaliyah	18.28a	14.16a	27.40a	21.73a
Mutla	20.12a	31.52a	34.10a	21.10a
Minagish	26.17a	15.30b	31.96a	15.56b
Maqwa	—	—	56.70a	32.00b
Ali Al-Salim Base	—	—	53.93a	46.30a
A. Al-Jabir Base	—	—	42.80a	21.37b
Hoban	—	—	55.60a	42.53b
Sulaibiya T. Station	—	—	62.96a	18.63b
Rawdatain	11.50a	14.15a	36.90a	29.36b
Sulaibiya (KISR)	43.20a	0.00b	59.00a	0.00b

*Means followed by the same letter in the same row in the same year are not significantly different at $P=0.05$ (Duncan's Multiple Range Test). — =no data taken.

Table 3. Production in kg/ha at each study location in 1980–1982.

		Production (kg)ha)							
		Forbs		*Grasses and grasslike*		*Shrubs*		*Total production*	
Location	*Year*	*I*	*O*	*I*	*O*	*I*	*O*	*I*	*O*
Abdaliyah	1980	1.85a	10.00a	42.59a	45.93a	59.73a	13.90a	104.18a	69.83b
	1981	25.80a	13.54a	54.76a	61.37a	235.17a	67.10a	315.73a	142.01b
	1982	0.00a	0.00a	17.78a	19.26a	7.83a	1.95a	25.61a	21.21a
Ali Al-Salim Base	1980	—	—	—	—	—	—	—	—
	1981	136.41a	118.63a	23.70a	27.63a	93.70a	186.77a	253.81a	333.03a
	1982	—	—	—	—	—	—	—	—
A. Al-Jabir Base	1980	—	—	—	—	—	—	—	—
	1981	120.00a	98.89a	2.44a	13.96b	182.92a	2.17b	305.36a	115.02b
	1982	—	—	—	—	—	—	—	—
Doha	1980	—	—	—	—	—	—	—	—
	1981	320.15a	41.76a	45.69a	68.46a	272.62a	38.20b	638.46a	148.42b
	1982	0.00a	0.00a	0.00a	0.00a	45.58a	2.17a	45.58a	2.17b
Hoban Tr. Station	1980	—	—	—	—	—	—	—	—
	1981	110.31a	30.80b	79.50a	114.11a	109.70a	31.64a	299.51a	176.55b
	1982	70.22a	0.00b	21.70a	17.13a	44.84a	9.57b	136.76a	26.70b
Minagish	1980	52.60a	126.30b	52.22a	50.74a	0.57a	1.23a	105.39a	178.27b
	1981	177.02a	108.63b	37.54a	16.33b	0.00a	16.94a	214.55a	141.90b
	1982	0.00a	0.00a	0.00a	0.00a	0.00a	3.38a	0.00a	3.38b
Maqwa	1980	—	—	—	—	—	—	—	—
	1981	180.81a	59.48b	15.87a	0.00b	66.71a	0.00b	263.39a	59.48b
	1982	294.32a	0.00b	0.00a	0.00a	20.29a	0.00a	314.61a	0.00b
Mutla	1980	53.70a	255.19b	65.74a	10.74b	9.83a	13.63a	129.27a	249.56b
	1981	—	—	—	—	—	—	—	—
Rawdatain	1980	119.26a	43.33a	73.70a	85.93a	54.60a	24.40a	247.56a	153.66b
	1981	220.96a	134.19b	83.39a	204.67a	73.09a	107.21a	377.44a	446.07a
	1982	0.00a	0.00a	0.00a	0.00a	3.25a	8.08a	3.25a	8.08a
Um-Ghara	1980	110.19a	84.44a	76.30a	32.22a	0.13a	8.70a	186.63a	125.36b
	1981	—	—	—	—	—	—	—	—
	1982	—	—	—	—	—	—	—	—
Sulaibiya Tr. Station	1980	—	—	—	—	—	—	—	—
	1981	150.20a	49.48b	5.94a	34.00b	95.43a	105.80a	251.57a	189.28a
	1982	9.38a	0.00b	16.05a	133.15b	14.79a	9.85a	40.21a	143.00b
KISR Sulaibiya	1980	308.81	—	36.95	—	105.65	—	451.41	—
	1981	185.70a	95.09b	9.89a	24.78a	96.99a	0.00b	291.58a	119.87b
	1982	56.80a	1.75b	91.91a	150.46a	34.48a	4.74b	183.19a	156.95a

* Means followed by the same letter in the same row in the same year are not significant at $P = 0.05$ (Duncan's Multiple Range Test). I = inside, O = outside

are highly degraded because of overgrazing, the rangelands have not yet reached the point of irreversibility in regard to forage improvement.

Annual herbage production

Production data for all the study sites were conducted for a three year period (1980, 1981 and 1982). Herbage production (kg/ha) for the vegetational classes (forbs, grasses and shrubs) at each study location is shown in Table 3. The data represent herbage production at peak biomass for the herbaceous as well as the shrubby species. Due to climate fluctuations between years, times of species harvest were made flexible to allow for clipping during peak biomass. Species for an individual vegetation class were individually clipped but were later combined for estimation of herbage production. Data were obtained inside and outside the fences of the 12 study locations. It should be mentioned here that the study locations at Abdaliya, Minagish, Mutla, Rawdatain, Um-Ghara and Sulaibiya Transmission Station were, at times, subjected to grazing pressures from break-in by nomads' animals.

Table 3 presents production of different classes of vegetation inside and outside exclosures during 1980, 1981 and 1982 for the 12 study locations. Forbs varied in 1980 from 10 kg/ha in Abdaliya to 308.8 kg/ha in KISR Sulaibiya. Forb production decreased to a significant value for all locations outside exclosures. This indicates the positive effect of fencing on herbage production. In 1981 the value ranged from 25 kg/ha in Abdaliya to 320 kg/ha in Doha. In 1982 all values declined due to the scarce amount of precipitation during that year. Grasses and shrubs were much lower in herbage production than forbs. This was encountered inside and outside exclosures.

Production inside exclosures for 1980 varied among sites, with KISR Sulaibiya producing 451.4 kg/ha and Minagish having only 105.3 kg/ha (Table 3). Average production for all sites was 205.4 kg/ha. In 1981, production inside Doha (salt marsh site) averaged 638.4 and Minagish only 214 kg/ha. Herbage production was generally high for all sites during that year compared to production in other years during the study period. In 1982, production varied between 314.6 kg/ha in Maqwa and 0 kg/ha in Minagish.

Overall average herbage production for the three-year study period is 206.7 kg/ha (dry matter). The variation encounterd during this period ranged from 321.1 to 933.6 kg/ha, which is significant enough to merit continuation of data collection until a realistic average for herbage production is obtained. Once this is achieved, it will be possible to estimate appropriate stocking rates and carrying capacity for the rangelands.

Phenology

Phenology is the morphological and physiological expression of a plant to seasonal changes in its physical environment. In order to live in a given location, the plant must be able to make full use of favourable environmental conditions and tolerate the unfavourable ones.

A three year record of the average phenophase description for all species was taken from the 15 study sites within the KISR Sulaibiya Station (Fig. 4). Phenological recording of species at individual sites was carried out. However, in order to simplify data interpretation, only the primary species will be discussed in this paper. In general terms, major phenological data for shrubs and forbs were as follows: vegetative growth varied from 10 January to 15 February, full bloom from 16 March to 30 March, and seed dissemination from 16 April to 30 May. As for grasses, the sequence was as follows: full vegetative growth varied from 13 January to 24 February, full bloom (anthesis) from 23 March to 20 April and seed dissemination from 5 April to the end of the growing season (June). When these data are examined, it becomes apparent that shrubs generally grew slowly through much of the growing season, reaching maturity in May. These species are thus able to provide much green forage for most of the grazing season. On the other hand, forbs which exhibit growth similar to that of shrubs, differed markedly in phenological dates and followed those of grasses closely. Perennial forbs grew for longer periods than grasses but fell short of the shrubs' long vegetative cycle. Conversely, annual forb species grew rapidly and completed their life cycle and thus provided forage for only a short period. The growth pattern for annual and perennial grasses was very similar. The perennial species, however, remained vegetative for longer periods and completed their life cycle later in the season.

The phenological progression of *Rhanterium epapposum* and *Cyperus conglomeratus* in relation to precipitation and average temperature during the study period cycles is shown in Figs 5 and 6. During 1979–1980 and 1980–1981 the first effective rainfall started in December (50 and 41 mm, respectively) and the two plants initiated vegetative growth in late December/early January. In 1981–1982, however, the first effective rainfall

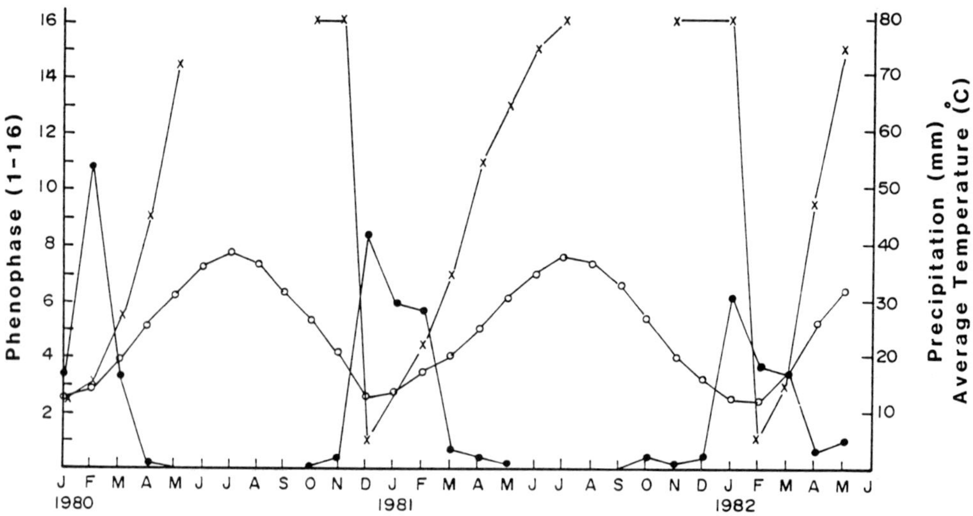

Fig. 5. Phenological progression of *Rhanterium epapposum* at KISR Sulaibiya Station in relation to precipitation and average temperature in a three-year cycle. ●——● Precipitation, ○——○ Temperature, ×——× Phenophase.

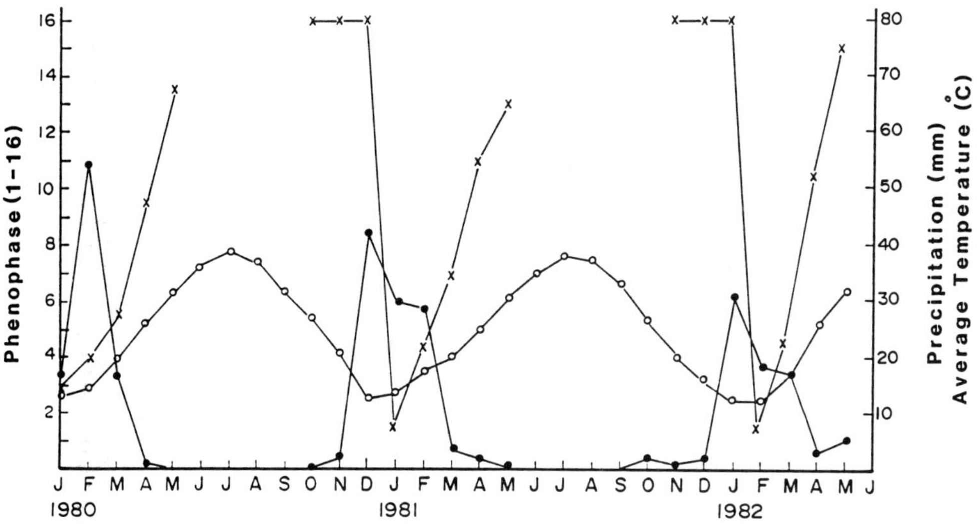

Fig. 6. Phenological progression of *Cyperus conglomeratus* at KISR Sulaibiya Station in relation to precipitation and average temperature in a three-year cycle. ●——● Precipitation, ○——○ Temperature, ×——× Phenophase.

started in January; consequently both plants initiated vegetative growth in February. The preliminary data gained from the phenological studies indicate the importance of the precipitation parameter and the timing of the first effective rainfall in the initiation of vegetative growth of the prime species.

Data on phenological progression of the prime species in relation to site parameter were collected at 15 subsites at KISR Sulaibiya Station. Results of the study were reported in detail (Taha *et al.* 1980). The data showed the effect of site variation on the phenophase description, i.e. different phenophases were encountered at different sites on the same date. Such variations can be attributed to differences in soil characteristics, topography, slope and exposure. The physical factors possibly affect phenophase stages through modifications of the earlyness and lateness of vegetative and reproductive stages (Caprico 1966, Fisser *et al.* 1976, 1977).

Qualitative evaluation of vegetation

The qualitative evaluation of vegetation was investigated by collecting range plants from the various vegetation and ecological zones and preparing baseline information about the plants of Kuwait's rangelands.

Extensive field trips were made to select field sites for plant collection. A list of all species present at each site was made. Plant specimens collected were mounted and identified. They were then placed in a newly established herbarium. The older herbarium of the Department of Agriculture, Agricultural Affairs and Fish Resources Authority, houses some 200 specimens which date back to 1960's. Specimens were reviewed and nomenclature verified as necessary. Based on the specimens at the two herbaria, a list of native plants in Kuwait was prepared.

SUMMARY AND CONCLUSIONS

An extensive vegetation survey was conducted at the KISR Sulaibiya site as well as at several other desert exclosures (Rawdatain, Mutla, Doha, Um-Ghara, Abdaliyah, Minagish, Hoban Transmission Station, Maqwa, Sulaibiya Transmission Station, Ali Al-Salim Base and Ahmad Al-Jabir Base). Vegetation measurements included percent cover, frequency, density, phenology and production. These parameters were all measured at these study locations, and the data obtained have provided impetus for comprehensive baseline information for management and improvement of Kuwait's rangelands. The study locations that enjoyed full protection from grazing have clearly demonstrated the significance of plant protection in terms of increased plant cover, frequency, density and production. This was demonstrated in the well protected exclosures at Ali Al-Salim Base, Sulaibiya Transmission Station, Hoban Transmission Station, Maqwa and Doha. Increase in plant cover inside the fenced areas varied between 63% (Sulaibiya Transmission Station) and 18% (Doha). Another interesting finding is that dominant species were found to respond favourably to protection, even if grazed periodically (Sulaibiya Transmission Station is a good example where light periodic grazing is practiced). *Rhanterium epapposum* and *Cyperus conglomeratus* increased significantly inside exclosures compared to the outside and despite Kuwait's harsh environment.

Annual herbage production has varied considerably among sites as well as among years. Average herbage production for 1980, 1981 and 1982 was respectively 205.4, 321.1 and 93.6 kg/ha (dry matter). These differences were attributed to variations in the amount and distribution of rainfall during these years. These greater variations among years in herbage production make it extremely difficult to draw a realistic overall estimate for herbage production in Kuwait's rangelands. It should be mentioned, however, that the data obtained indicate that, despite the known fact that Kuwait's rangelands are highly degraded because of overgrazing, the rangelands have not yet reached the point of irreversibility and thus there exists some potential for improvement.

Recording the chronological dates for the various developmental stages of plant growth for most range vegetation was carried out at KISR Sulaibiya Station. Major events of the plant life cycle such as initial leaf growth, full leaf growth, flowering, seed maturation and seed dissemination were identified throughout the three-year study period. Beause of climatic fluctuations in terms of precipitation and, to some extent, temperature, however, the plants exhibited greater variations in growth pattern than expected. Such variations in a chronological record of species' developmental stages indicate the need for continuation of such studies until certain patterns are evolved for the major events of the plant life cycle. Data have also clearly demonstrated that physical factors such as soil type and texture, topography, relief and exposure either enhanced or delayed initial plant growth or hastened or slowed their maturity. Future plans call for more intensive field observations as well as detailed analyses of physical and microclimatic parameters. Once these data are completed, it will be possible to tailor grazing and range improvement practices according to phenological dates.

Another aspect of the vegetation survey was species collection and herbarium organization. Extensive field trips were conducted throughout Kuwait's rangelands to collect, identify and mount native species. As a result, the nucleus for a KISR Herbarium is now established with more than 200 species.

LITERATURE CITED

Bennachio, S. S. 1970. Phenological development of some perennial species as related to environmental factors in Indiana. Ph.D. Dissertation, Purdue University, Lafayette, Indiana.

Brown, Dorothy. 1954. *Methods of measuring vegetation.* Commonwealth Agri. Bureau, Farnham Rl, Bucks, England.

Caprico, J. M. 1966. Pattern of plant development in the western United States. *Montana Agri. Exp. Sta.* Bull. 607.

Carnfield, R. H. 1941. Application of the line interception method in sampling range vegetation. *J. Forest.* **39:** 388–394.

Dyksterhuis, E. J. 1949. Condition and management of rangeland based on quantitative ecology. *J. Range Manage.* **2:** 438–444.

Dyksterhuis, E. J. 1958. Ecological principles in range evaluation. *Bot. Rev.* **24** (5): 253–272.

Ergun, H. 1969. Reconnaissance soil survey. Report to the Government of Kuwait. FAO/KU/TF17. Mimeo, Rome.

Fisser, H. G., Jerry Cox and Faisal K. Taha. 1976. Herbage structure, production and phenology monitoring on the Black Thunder. *Wyoming Agr. Exp. Sta.* Sci. Rep. 779.

Fisser, H. G., Jerry Cox and Faisal K. Taha. 1977. Herbage structure, production and phenology monitoring on the Black Thunder. *Wyoming Agri. Exp. Sta.* Sci. Rep. 832.

Halwagy, R. 1986. On the ecology and vegetation of Kuwait. p. 81–109. *In:* Kürschner, H. (ed), *Contribution to the vegetation of Southwest Asia.* Dr Ludwig Reichert Verlag, Wiesbaden, West Germany.

Halwagy, R. and M. Halwagy. 1974. Ecological studies on the desert of Kuwait. (i) The Physical Environment. (ii) The Vegetation. *J. Univ. Kuwait (Sci.)* **1:** 75–95.

Parker, J. F. and G. D. Pickford. 1937. A weight estimate method for the determination of range and pasture production. *J. Amer. Soc. Agron.* **29:** 894–904.

Taha, F. K., S. A. Omar, M. Sartawi and G. Nakshabandi. 1980. Advances in range management in Kuwait. Kuwait Institute for Scientific Research, KISR Tech. Rep. No. 253.

Improvement of arid rangeland

HAROLD F. HEADY

145 Mulford Hall, University of California, Berkeley, California 94720, USA

ABSTRACT

Deteriorated rangelands in arid regions, particularly those in developing countries, are complex as to the causes of their poor condition, and recommendations to improve them are usually over-simplified. How should improvement start when the pastoralists are living near poverty? The answer must be partly social, so establishment of range management cooperatives is suggested as a means of gaining the confidence of the people and serving as an educational center. An item of major concern is the goal of producing more forage rather than emphasis on reducing livestock numbers. This may be accomplished through increased production of forages on the small farms and better integration of crops, herbaceous forages, shrubs and trees. Proper management of water is crucial to grazing management. Proper distribution of water is required for rotation of grazing. Ungrazed reserves show the region's potential for vegetational cover as well as offering places for beauty and the pleasure of people. All rangeland practices for better management must be altered to fit local conditions, regardless of where they originated.

INTRODUCTION

Both rangeland deterioration and improvement are complex processes; the existing situations being the products of social changes, human population shifts, fluctuation in numbers of domestic and wild animals increasing economic pressures, advancements in medicine for people and animals, fluctuations in climate, the careless replacement of tradition with undesirable consequences and others. Each of these subjects is worthy of full discussion in the examination of rangeland improvement. Consideration of their interactions with each other could well be of greater importance than analysis of individual factors, but these may not be very profitable until more is known of the individual factors. The complexity of interactions, lack of knowledge of all the factors, and shortage of allotted time forces me to restrict this paper to a few topics, mostly the physical rangeland improvements.

However, all of us should keep in mind that people and their animals evolved in rangeland ecosystems. Many still live in those systems and others, who do not, depend to some degree on the goods and services that rangelands provide. Those who reside in a rangeland environment obviously gain from the residence. They also influence or service the rangeland ecosystems and that is the greatest concern in rangeland improvements.

Recommendation of range improvements in desert areas is oversimplified in most situations. Governments become aware of deterioration of vegetation and soil and rush to settle people at any cost. The goat is singled out and maligned for destroying any trees and shrubs that remain. As populations of people and animals increase, so does the pressure on national and international agencies to develop water, but with control neither of the use of water nor of the surrounding rangeland. The concern is real but the answers are more fundamental.

How should improvement start?

Always, the first question should be: "How do you start from where people are in attaining rangeland improvement?" That question is doubly more troublesome for dry climates than for humid regions where the grasses are tall and grow rapidly. A return to few people on the land and widespread nomadism is no more desirable than it is feasible. Somehow the modern technology of rangeland management must be made applicable to the land and to the lives of peoples in developing countries, much of which is arid and semiarid and most of the people near poverty. Following are some ideas and suggestions about the application of rangeland improvement in dry climates between 30°N and south of the equator and with special thought to developing countries.

Regional management through grazing cooperatives

The purpose of the grazing cooperatives is to modernize extensive grazing. But first there must be an undertanding of the indigenous people. Regional management implies leadership, organization, and rules made and applied by the users of the rangeland. Understanding of the human and traditional base is the starting point. Such items as numbers of animals, migratory movements, security against animal loss, settlement of landuse conflicts, making investments in the land, and periods of grazing are traditional functions of the tribe or family group as a whole and are enforced by its leadership. Each is a starting point for reorganizing landuse practices. Beginning with traditional management encourages participation by the beneficiaries. Little can be accomplished without that cooperation.

Cooperative organizations are groups of people organized for their common concerns and they are managed by their own members. They constitute centers for social affairs, for economic reasons, as well as for technical matters. A cooperative is a device for involvement of people.

The advisor who establishes one learns in its organization. After a cooperative is established, the technical advice of range management specialists, extension and training personnel, and aid groups from the central government have a place where they explain and make their services available to the people. Many different kinds of physical range improvements can be established. For example, after the 1974 drought when rainfall returned to normal in Niger, excessive pressure remained on the rangelands through the rapid increase in livestock numbers and new agricultural settlements. However, the cooperatives, usually around pumping stations, were the centers for full agricultural education and cooperative effort for better land management, although not always successful.

Livestock numbers and grazing pressure

Reduction of herd size is a universal cry for reversing rangeland deterioration, desertification, and the periodic ravaging by drought. There is no question that overgrazing has and still occurs, but in very few situations have livestock numbers been reduced in developing areas, except through animal deaths in drought times. Neither understood nor accepted is the fact that the first half of full feed consumption by grazing animals goes to maintenance of the animal body. The second half is used for reproduction, milk, growth and fat storage, none of which occur if the animal is on a half or less diet. Fewer animals means more feed in the second half of the diet and more animal products for people.

Much has been written on the reasons why these aspects of animal husbandry have not been accepted and schemes to reduce livestock numbers have failed. Technical people tend to equate livestock production with beef production and want fewer livestock numbers; but milk is probably the most important animal product in developing countries and it is needed every day. The facts that most cows in the developing areas freshen but once every three years and provide little milk in the dry season are reasons why large numbers of animals must be maintained. We do not accept that a 30% calf crop is the goal, but will not discuss the animal breeding issues involved. To solve the problem something more is required. One of the possibilities is the milking of more sheep and goats. Another is to increase the feed supply. Forage productivity, then, is the thrust and the aim is a balance between the herd itself and the producing of forages. Hopefully, this will lead to reasonable grazing pressure without the stigma of being labelled in the unfavourable words of fewer animals.

Improved production on the small farm

In general terms, the land-dwelling population of many countries operates in three systems: one is river- or oasis-based depending on the use of water for cultivation of foods and forages. Rainfed cropping, generally restricted to areas receiving more than 400 mm precipitation per year, is of this type and the population is sedentary. The second is extensive rangeland grazing by a nomadic population. The third is a mixture of the two and is based on a combination of crops, extensive rangeland grazing, cropland grazing, and sometimes fodder and food trees. Although nomadic grazing occupies the larger area, more livestock occur in the sedentary systems.

Many changes have occurred in these systems in the last 30 years. Nomadism has been curtailed through the establishment of schools, new water points, dispensaries, and enlarged communication via roads, vehicles, and radio. Sedentarization has resulted in more herds of smaller size and increased overgrazing near water points. Increase in the area farmed often has been at the expense of the best rangeland. Local erosion has increased (UNEP 1977). These changes require that rangeland management not be an end in itself. Although sedentarization of livestock and cropping are irreversible and the dry areas can only be used for extensive grazing, many interrelationships exist between the rangelands and the croplands. The irregularity of that range resource can be met in part by integrated use of feeds from cultivated and range areas. A policy and development aimed toward the improved integration of rangelands with cropping will improve the lives of rural peoples, farmer and pastoralist alike.

The village farmer with a hectare of land and a few animals may well be the key to the improved management of rangeland. For example, animals maintained on rangeland during the wet season could use the crop aftermath in the dry season. Balancing numbers of animals with the yearlong forages is critical. Progress in all phases of this type of production system is essential for increased food, employment, and income. Most research worldwide and most development schemes have aimed at large scale monocultures for exchange in large, often foreign, markets and for increasing the gross national income. People on the small farms have not shared equally in this technological revolution whether it be for grains, animal products, or wood products. The fact that cotton from large irrigation schemes has received more attention than millet for family consumption and local trade, illustrates the lack of attention to the rural people themselves.

Much of the rangeland overgrazing in developing areas lies near the villages. Sometimes the few animals owned by each person are herded together, but that is of little consequence. What matters is that every day all the animals go to and from the nearby rangeland resulting in heavy grazing pressure on the vegetation. The worst overgrazing occurs close to the villages. As the villages become larger, so do the areas cultivated, the number of animals increases, and the overgazing becomes more severe. Improved farming in the villages is crucial to improved rangeland management. One cannot be attained without the other. Oasis crops always produce feed for animals including items like olive cuttings, date by-products, sesame cake, and *Medicago* grown for cut-and-carry feeding. These need attention from the scientist and manager in order to do more for the village farmers and their herds. Marginal land for cropping should be converted to improved pastures and agroforestry uses.

Water development and management

Although discussed in other papers, water for people and animals is crucial for the development of arid lands and must be mentioned in this paper on rangeland improvement. Providing water usually shows more visible results than other rangeland improvements. The placing and capacity of water developments are manageable, but often little attention is given to these specifications. When water becomes available, concentrations of people and animals tend to destroy the resources. A water survey in Botswana (Roe 1984) suggested that restricted access, especially in the wet season, is a way to maintain reasonable range condition nearby.

In many rangeland and animal intervention schemes and frequently at country level, water policy gives priority to helping herds during the dry season. Management of water and rangeland near water in the wet season has received little attention. The critical period for the vegetation is during the growing season, and it must not be overgrazed at that time.

Increased production of feed

Revegetation includes a group of practices aimed at restoring the productivity of the land to its full potential. Management is the least costly and easiest route to follow and the result is natural vegetation of the type expected for the site. It should be the initial step. If that goal cannot be reached through natural plant succession as encouraged by animal management, then the recourse is to planting and seeding. They should begin on suitable sites where the vegetation had not responded to management, but should not be attempted where grazing cannot be eliminated for the establishment period and managed thereafter.

Experience has shown that species must be adapted to the site and environment. Few are available for arid conditions but there is interest in *Acacia*, *Atriplex*, cactus, and other shrubs. Legumes are highly desirable for rangeland seeding but few are suitable for arid lands. Effort should be taken to reduce plant competition to the seeded species, whatever the climate and site. Planting according to specified time, seedbed preparation, and depth are important for success.

Accumulation of information on the culture and characteristics of a number of the desert plants would appear to be worthwhile. In particular, species of *Salsola*, *Artemisia*, *Haloxylon*, and *Zygophyllum* should be studied for their properties of furnishing forage, wood, and sand control. The wide spread *Panicum turgidum* has received little attention as a possibility for rangeland rehabilitation. A few legumes (*Stylosanthes* and *Acacia*) and the chenopod, *Kochia* would be of interest. Few, if any, of the commonly used species for rangeland seeding in temperate areas appear worthwhile for testing in the tropical and subtropical dry areas. The available grasses all require 250 mm of rainfall and most of them much more. The most drought resistant of the developed subtropical legumes still require approximately 500 mm of rainfall.

Deserts, like all other lands, have a variety of sites or habitats. Sandy areas and pure sands may produce vegetation beyond expectations from their rainfall because there is no runoff. All the rainfall enters the sand, penetrates deeply, and is protected from deep evaporation by the dry surface layer. The flat wadi bottoms, many of which are sandy, receive run-on water from rainfall on the surrounding clay and stony lands. These sand rivers constitute reservoirs of soil and water that can produce useful plants beyond the average rainfall of the region. These and others of the very best sites are the ones that should receive the first and major attention in desert management. Thus, species of *Cenchrus* which require 400 mm or more of rainfall could be grown on favourable sites where the rainfall is less.

Agroforestry

Wherever environmental conditions are favourable to the growth and development of trees, their culture is always beneficial for wood, fence rows, shade, marking boundaries, fuel, soil enrichment and fodder; for creating employment and income; and for reducing erosion. Woody plants and grasses are used for the control of sand dunes. The concept of agroforestry emphasizes production of food crops, forage, and woody plants on the same land base. Values of trees and shrubs go far beyond fodder purposes.

Integrating woody plants into rangeland improvements is a current and worthwhile undertaking, but establishing and producing trees and shrubs are exceptionally difficult challenges in arid areas. Selected sites have possibilities and the search for adapted plants has just started. Success can be only through an interdisciplinary process that will bring together the scientists and professionals from agronomy, forestry, animal husbandry, wildlife, range management, and social science disciplines. Study of rangeland systems, peoples, animals, institutions, and interventions as described by Child *et al.* (1985) is appropriate as it discusses the problems of rangeland production in developing countries and provides information on agroforestry.

Seasonal grazing systems

Among other things, grazing management requires control over the defoliation of plants by animals, and physical damage to the plants by animal trampling and by pulling of individual grasses from the ground or parts from the woody plants. Plants react to these defoliations and other impacts in various ways. Some are stimulated to a high degree by grazing while other are damaged by even a minimum of foliage removal. Generally speaking, all range plants can be grazed to some degree at certain times in the growth cycle. The manager's purpose is grazing at the highest degree that still maintains improving range conditions and the highest production levels.

Grazing is a natural and normal process whether one is considerig dense grasslands or scattered desert shrubs in arid regions. Grazing, or more specifically defoliation, operates in several ways to alter the vegetation. One is the intensity to which it is done. This involves numbers of animals and the length of time that they graze an area. Management of animal numbers was mentioned earlier; regardless of the management employed, overgrazing, that is over-defoliation, is to be avoided.

A second aspect of defoliation is the frequency with which it occurs. Plants must have a certain amount of green leafage or stem during the growing period if they are to survive. Frequent defoliations that continually remove the green growth are especially damaging. In fact, frequent grazings tend to be damaging at any time in the growth cycle, if they prevent transfer and accumulation of carbohydrates from the new growth where they are manufactured to other parts of the plant.

A third aspect of defoliation is the selectivity of plants and plant parts by the grazing animals. It is well known that some species are selected before others and that different kinds of animals, even different animals, prefer various forages. No two diets are the same. When one plant is over-defoliated, it becomes weakened and at a disadvantage in the competition for light, minerals, water, and space with other plants. This leads to vegetational changes. If the declining species are the most desirable, the rangeland condition is declining. When carried to extreme, the final result is bare soil, erosion, and survival problems for plants, animals, and humans.

Still a fourth impact of defoliation is the season during the growth cycle when it occurs. Range managers know that every plant reacts differently to defoliation as the growth season progresses. Grazing in the dry season when plants are mature is less damaging for nearly all species than grazing in the wet season. Woody plants are more sensitive in this respect than the grasses. However, there is little information on the response of desert plants of the tropical and subtropical world. In temperate regions, much has been said in recent years about the use of seasonal grazing systems in order to manage the impact of defoliation at different times in the plant's growth cycle. It is a difficult problem because the animals must eat every day of the year so some plants must be grazed every day in the year. This means that some plants are grazed when they are most susceptible to damage. Herein lies the interest in seasonal grazing rotations for they ensure that individual plants will be grazed for only a limited time during the sensitive growth period. Further, the grazing at the sensitive time is rotated so that the pastures in turn have plant growth periods without grazing.

Fortunately, grazing is manageable, but with numerous constraints that are related to tradition, vegetation, environment, and animals. Grazing systems, as a part of range management, aim at overcoming those constraints. Among the more successful rangeland grazing schemes, including arid regions, are those in which separate and designated areas are grazed in some sort of rotation. The vegetation is more sensitive to defoliation at different times. The principle is to rotate the grazing pressure so that the same vegetation is not grazed repeatedly at the sensitive stage of growth during every growing season. Rotations have fostered improvement in rangeland condition in some types of range vegetation but not in all. Arid areas may show little change after several years without animals; so rapid results in improving range conditions of deserts are not to be expected from the use of seasonal grazing schedules.

Other types of systems include transhumance whereby all animals are taken to summer or winter pastures or to wet or dry season rangelands. The animals are then returned to rangelands near the permanent home and are grazed on cropland as well as local rangeland. The latter has been referred to as fallow and stubble grazing. Rotational grazing usually refers to some sort of scheduled transfer within the transhumant or the rangeland/cropland systems. A characteristic of these rotations is that they require multiplication of more or less permanent water, and abundant fencing or other kinds of marking of pasture boundaries. Rotations have had a measure of success without fencing where continuous herding of animals is an everyday practice, even from a very early time.

Draz (1983) described the *hema* system of grazing that was practiced several hundred years ago by provincial chiefs or other tribal authorities in the Arab world. The system protects the land in five ways: (1) complete prohibition of grazing but cutting of grass allowed at certain times and places; (2) grazing and cutting at certain times of the year; (3) yearlong grazing with kinds of animals and numbers specified; (4) no grazing until after flowering for bee-keeping purposes; (5) grazing restricted to protect trees. These are excellent principles for the management of rangelands. Of note is the fact that they came into use with some managerial and ecological understanding long before the physiological responses of plants had been researched. Also of note, the physiology of defoliation in its four major aspects is still not understood for many species. This is a challenge for ecophysiology of plants of arid lands.

Grazing reserves, parks, protected areas, wildlife refuges

These types of preserves give the natural ecosystems a chance to develop and thereby show the nature and density of the native plants and animals that will evolve without grazing by domestic animals. In most instances, these areas should not be grazed by domestic animals, but only by the wild animals. In developing countries of the arid and semiarid zone, the wild animals will need introduction, management and protection if they may have long since disappeared. Livestock-free areas eventually show the flexibility of rangeland ecosystems, the diversity of resources, and above all the basic soil resource at its best. Reserves of various kinds are for various purposes among which are the demonstration and education of the public as well as livestock owners on vegetational ecology under grazing and without it. Preserves are laboratories for study and teaching as to the nature of changes in unreasonably used resources and the characteristics of rangeland potentials with proper use. Kuwait is to be commended for the establishment of preserves that are large enough to give whole ecosystems opportunity to show that the desert is productive. Other countries with large blocks of arid and semiarid land should do the same.

LITERATURE CITED

Child, R. D., H. F. Heady, W. C. Hickey, R. A. Peterson and R. D. Pieper. 1985. Arid and semiarid lands: Sustainable use and management in developing countries. Winrock International, Morrilton, Arkansas, USA.

Draz, O. 1983. The Syrian Arab Republic rangeland conservation and development. *World Animal Rev.* **47:** 2–14.

Roe, E. M. 1984. Range conditions around water sources in Botswana and Kenya. *Rangelands* **6** (6): 247–249.

UNEP. 1977. Desertification: An Overview. United Nations Conference on Desertification. Nairobi. 79 pp.

FAO position paper on a regional range management and fodder production programme for the Gulf Cooperation Council Member Countries*

MAHGOUB G. ZAROUG†

Regional Management and Fodder Production Officer FAO/RNEA 00100 Rome, Italy

ABSTRACT

Member Countries of the Gulf Cooperation Council (GCC) have initiated various programmes and are undertaking appreciable efforts to expand and improve the efficiency of their agricultural production systems in order to secure food for their expanding population and to achieve betterment in the living conditions of rural communities. Being aware of the important role played by the range resource in providing livestock feed and protecting the soil from erosion, the authorities in these countries gave considerable attention to the conservation and development of this resource. Efforts were also made to develop fodder and pasture crops with the objective of reducing critical shortages in livestock feed.

This paper draws attention to major constraints on range, pasture and fodder crops and points out some possible approaches and practices to cope with the obstacles hindering the development of the forage resources. On rangelands, natural aridity, expanding livestock population and irrational distribution of water and subsidized feed are major constraints. Pasture and fodder crops are constrained by poor agronomic practices, lack of good quality seed, infestation by pests and infection by disease-causing organisms.

The paper suggests that the contribution of rangelands, cultivated pastures and fodder crops to total livestock feed resources can be increased: (i) for the range resource, by reviving traditional practices for range utilization after introducing necessary supporting amendments, by matching livestock population and length of grazing season to particular site abilities to produce forage, and by increasing forage supply from fodder crops, pastures and fodder shrubs to supplement naturally produced forage; (ii) for fodder crops and pastures, by increasing forage production through the improvement of agronomic practices such as the use of good quality seed,

*Bahrain, Kuwait, Qatar, Saudi Arabia, Sultanate of Oman and United Arab Emirates.

†Present address: Ministry of Agriculture, Salalah, Sultanate of Oman.

use of adaptable species and varieties, and the utilization of salt-tolerant forage species and relevant technology for cultivating salt-affected lands.

The role of research, demonstration and training in developing forage resources and in the dissemination of required material and appropriate technology is emphasized. The paper also points out to the immediate need to strengthen education/training programmes to benefit professionals, technicians and land users.

A possible strategy in the form of a network scheme to strengthen current and create new collaborative activities is defined. The components of this network can be effectively analyzed in a meeting attended by representatives of member countries and GCC secretariat in order to formulate objectives, define priority problems and elaborate methods and approaches. Although the proposed programme is regional, yet local variations pertinent to natural resources or socio-economic factors must be taken into account.

1. BACKGROUND

1.1 General

Countries of the Gulf Cooperation Council occupy the major part of the Arabian Peninsula with a total land area of 247.5 million hectares (2.48 million km^2). Climatologically, the whole area is classified as arid but there are localized pockets with distinct environmental conditions that differ from the major land mass in their climatic parameters and consequently resources and land use. Prominent among these are Asir Mountains in Saudi Arabia and the mountain-plateau system of the Southern Region of the Sultanate of Oman. The region has one of the harshest environments in the world, e.g. Rub Al-Khali (Empty Quarter) which consists of extensive sand dune formations. Rainfall is generally low, coming during the cool part of the year except in parts of southwest Saudi Arabia and the Southern Region of Oman where summer rain and mist occur. Arising from the arid nature of the region and the absence of perennial streams or rivers, water is the most important single resource limiting agricultural development. Groundwater is important for supplying the needs of irrigated gardens and farmland. Seasonal drainage channels and wadis may flow following erratic rainfall. Their water recharges the subterranean aquifers, and a variable proportion flows into the sea or the interior desert. The region comprises various geological formations and types of soils dominated by aridic or desert soils.

In general, the erratic rainfall and the long rainless summer largely limit species of plants that can grow in deserts and coastal areas. Both of these habitats are characterized by the presence of extensive salt-affected lands dominated by halophytic vegetation. The desert areas have sparse vegetation cover consisting mainly of drought resistant shrubs supplemented by a range of perennial, annual and ephemeral grasses and forbs. Wadis and drainage channels and some basins subjected to runoff are relatively more vegetated and can support shrub and tree species. Likewise, mountains which receive sufficient precipitation are relatively vegetated. However, a large part of the tree and shrub component has been cut to meet wood and fuel requirements of local communities.

1.2 Rangelands

The region has a long history of pastoral activities and livestock raising used to be one of the major occupations of its people. Migrating flocks of sheep and goats and herds of camels move to utilize available forage following occurrence of rainfall and flow of major wadi systems and drainage channels. Desert range is used mainly during the cool part of the year while vegetated wadis and animal feed resources of the scattered oases are used during the summer. Because of animal movements between desert, wadis and oases, and due to the small number of animals, range use was within its potential forage production and capacity to support herds and flocks. Settled flocks utilize forage available in the immediate vicinity of villages supplemented by weedy growth in the irrigated gardens, fodder and other farm by-products. In the unique grasslands of the Southern Region of the Sultanate of Oman, there are two types of livestock migration. Cattle move from the mountain-plateau system to the foot hills to utilize the short grass that grows during the rainy (monsoon) season. They head back to the mountain-plateau system at the end of the rainy season, thereby deferring grazing on the cool and dry season areas and avoiding the irritating biting flies that are common during the monsoons. Camels and goats migrate to the south to utilize the short grass association of the Salalah Plain or to the north to use the desert scrub vegetation of the Njed, thus avoiding the biting flies during the rainy season. At the end of the monsoons they move to the escarpements and steep-sided wadis to make use of the available browse and other vegetation. Although this latter pattern of movement is continuing to the present time, yet the traditional movement of cattle has completely stopped; however, cattle move within the mountain-plateau system in search of natural forage.

There are now indications that the rangelands of the region are deteriorating at variable rates due to: (i) excessive grazing particularly around permanent settlements and perennial watering stations; (ii) distribution of subsidized feed which upsets the traditional movement of livestock that used to allow range vegetation to rest from grazing and trampling; and (iii) uprooting and cutting of shrubs for fuel or irrigated farming. On some sites, desirable species could dominate again or at least increase in abundance with proper management, while in many others assisted regeneration is required.

Both sedentary and migratory livestock face shortage of feed. Continuous use of grazing areas immediately surrounding villages results in vegetation destruction. The range resource is on the downward trend and loss of palatable species forces animals to consume all or part of plants which may contain potentially toxic compounds. Some complaints from livestock owners about the adverse effects of such plants on their animals were reported but several other cases were not. Species like *Arnebia hispidissima* (a relative of *A. nobilis* known to contain compounds which have medicinal use and violet colouration), *Zygophyllum simplex* and *Z. dumosum* (normally nitrite accumulators), *Crotalaria wissmanii* and *Tephrosia apollinea* (toxic amino acids), may cause digestive disturbances and other physiological disorders and there may be death in some cases.

There is a need to recognize that due to aridity of the region and the wide fluctuations in both amount and distribution of rainfall, native forge production from rangelands falls short of meeting the requirements of an expanding livestock population on year-long basis. It is therefore necessary to control livestock populations so that the general overgrazing is reduced. Herds and flocks should be distributed to make use of the forage produced in distant areas during the short rainy season in order to release grazing pressure from the degraded rangelands.

The importance of rangelands in the region is not solely attributable to the natural forage resource utilized by livestock but also due to the role of their vegetation in protecting soil and watershed, and contributing to the efficient harvesting of rainfall. This role is most appreciated when consideration is given to the fact that there are common watersheds and subterranean aquifers between countries of the region. Misuse of range vegetation in one area may translate into loss of valuable water resource and reduction in groundwater recharge which can ultimately influence water users in neighbouring areas.

1.3 Fodder crops

Fodder crops are traditionally grown in irrigated gardens in oases and coastal areas depending upon the availability of groundwater. Common among these crops are alfalfa, barley, sorghum, cowpeas and millet. The latter three are grown as grain crops under rain-fed agriculture in small holdings of the mountain-plateau system of the Southern Region of the Sultanate of Oman. Their stubble is used as animal feed. Alfalfa is grown as pure stand but sometimes this crop as well as millet are inter-planted with barley or grown on the boundaries of basins and irrigation channels. Barley is grown alone or mixed with alfalfa to establish a new stand of the perennial legume. A mixed grass-legume forage is obtained from the first cutting. Fodder crop yields are generally low due to various factors, the most important of which are: (i) poorly prepared seed-bed; (ii) use of low-quality seed; (iii) inappropriate irrigation water application and management; (iv) inefficient weed control; (v) pest and disease problems; and (vi) lack of or inadequate fertilizer application.

Large scale animal production farms established in the different countries of the region depend on irrigated fodder crops as a source of roughage. In general, these farms apply better agronomic and management practices, e.g. proper land preparation, fertilizer application, relatively more efficient irrigation (central pivots, dolphin guns, side-rolls and portable sprinklers). Fodder crops include multi-cut sorghum, sorghum × Sudan grass hybrids, maize, Rhodes grass and alfalfa varieties. The forage produced is used as green chop, baled hay or silage. Realizing the high demand for forage for the production of fresh milk, meat and supplementation of livestock and the high prices of roughage in local markets, more attention should be given to increase the economic efficiency of these large scale units through judicious use of inputs and high level of management.

1.4 Livestock

The livestock population of the region comprises camels, sheep and goats distributed in desert and coastal areas and to some extent in mountain regions. Cattle are confined to irrigated farming areas where supplemental green forage, weeds and stubble are available. An exception to this generalization is the cattle population raised in the grasslands of the Southern Region of Oman where natural forage resources form the major source of feed. Horses, mules and donkeys are present in the region but will not be dealt with in this report. Table 1 shows the population of these livestock in the countries of the region. Most of these animals are kept under traditional systems of nomadic, semi-nomadic and sedentary livestock. Cattle, sheep and goats have undergone large increases in the past few years (Table 2). Camels are reported to be declining in some countries such as Bahrain, Kuwait and Qatar while they have increased in others, e.g. Oman.

In their meeting in Riyadh during February 1986, GCC Ministers of Agriculture recommended the conservation and improvement of local types of animals particularly the Arabian horse, camels, goats and sheep. In many instances adequate feed supplies constitute a major constraint on traditional livestock with the result that local types cannot express their potential for

Table 1. Livestock population* (000) of the GCC member countries.

	Camels	*Cattle*	*Sheep*	*Goats*
Bahrain	NA**	6	7	15
Kuwait	5	18	550	300
Oman	6***	150	140	250***
Qatar	6	10	53	57
Saudi Arabia	160	500	3,500	2,300
United Arab Emirates	70	30	140	400

*Source 1983 FAO Production Yearbook.

**Not available.

***GRM International Livestock Inventory (1982) indicated that camel population in Oman is in the vicinity of 72,000 heads and goat population is about 700,000 heads.

Table 2. Relative* increase in cattle, sheep and goats in GCC member countries.

	Cattle	*Sheep*	*Goats*
Bahrain	150	175	188
Kuwait	257	550	375
Oman**	158	356	195
Qatar	167	136	127
Saudi Arabia	161	116	135
United Arab Emirates	NA	NA	NA

*Livestock populations of 1983 expressed as percentage of 1974 populations.

**FAO Production Yearbook data are not available for 1974 and other estimates were used.

production and reproduction. Total animal feed resources need to be increased to enhance the contribution of local herds and flocks to the subsistence of rural communities and to the production of surplus that can be utilized by urban populations.

Modern livestock production farms raising cross-bed (exotic × local) animals are being established for the production of milk but also to produce sheep and goat for slaughtering. Modern intensive livestock farms development can be encouraged but should not obscure vision of the need to assist traditional animal production systems so that they can effectively play their essential role in utilizing arid rangelands and converting the seasonally available forage into products highly needed in the region.

1.5 Recent developments and activities related to range, pastures and fodder crops

Rangelands are an important resource in the region, covering 35% of the total land area, as compared to less than 1 and 0.5% for forest-woodland and arable land respectively (FAO Production Yearbook 1983). There is an increasing awareness of the value of the range resource and the need to conserve it and develop it for supplying livestock feed and protecting the soil. In pursuance of those goals, countries of the GCC have initiated various activities related to surveying rangelands, range management, pastures and fodder crops production. These activities have been undertaken as part of national institutions programmes or through internationally assisted projects and collaborative work.

In Bahrain trials were conducted on production of irrigated fodder for sheep, goats and dairy cattle at Um-El-Nukheila and Hamala Stations. The Animal Production and Health Project (BAH/78/003) introduced elephant grass to farms in order to increase forage production and improve the ration offered to livestock by mixing forage from legume (alfalfa) and grass (elephant grass). Results were reported to be positive with very high yield from elephant grass during the hot season from May to September. Other promising fodder crops included multi-cut hybrid sorghum and Rhodes grass. These introductions improved forage quantity and quality with a desirable impact on dairy production and on raising lambs for slaughter.

Range and feed resource surveys in Kuwait were carried out by FAO, ACSAD (Arab Center for Studies of Arid and Drylands) and AOAD (Arab Organization for Agricultural Development) in order to assess available resources and commence some programmes for their development. Kernick (1966) described range vegetation in Kuwait and provided measurements about range cover and trends in the plant communities of that arid environment. Trials were conducted on introduced grass, legume and shrub species to assess their adaptability and potential under Kuwait conditions. From ecological studies in protected enclosures, the ability of range vegetation to produce a reasonable amount of forage, even in relatively dry seasons, was demonstrated.

Range and livestock development activities of KISR (Kuwait Institute for Scientific Research) are contributing to the integration of forage resources from rangelands and irrigated fodder and pasture species. The aim of these activities is to increase both animal feed resources and livestock products. Early results reported by Taha *et al.* (1984) were encouraging.

In Qatar, range survey and taxonomic studies of the country's flora were carried out. Trials on adaptation of *Acacia cyanophylla*, *Atriplex nummularia*, *Chloris gayana* and *Pennisetum purpureum* under irrigation were conducted in Abu-Samarah Sheep Farm. Agronomic trials on several varieties of barley, triticale (both salt tolerant), sorghum and alfalfa were also reported.

Rangeland conservation and development programmes in Saudi Arabia comprise range resource surveys, range improvement through combined seeding and water spreading, range vegetation studies and development of practical management systems. Recently, the Range and Animal Development and Research Centre has been established at Al-Jouf District to carry out research, and extension and training programmes in an important range area of the country. Experimental work on irrigated fodder crops, pastures and fodder shrubs is being conducted at Kuneitra and Tabarjal. Several trials on response of Rhodes grass, winter cereal forages and sorghum to nitrogen or to nitrogen and iron were carried out as part of the Joint Agricultural Research and Development Project which is a collaborative programme between the Ministry of Agriculture and Water in Saudi Arabia and the University College of North Wales. Some trials were also done on response of introduced and local alfalfa varieties to different irrigation regimes, fertilizer application and other agronomic practices.

Resource surveys for the rangelands in the Sultanate of Oman were conducted by AOAD, ACSAD and by the Ministry of Agriculture and Fisheries. Seeding and range fertilization trials carried out by IRI (the International Research Institute) in the Southern Region showed that active forage was increased by 66 to 200% when N and NPK were applied respectively. Seeding trials using the tropical grasses *Cenchrus ciliaris*, *Chloris gayana* and *Panicum antidotale*, and the legumes *Macroptillium atropurpureum*, *Glycine wightii* and *Stylosanthes* spp. have demonstrated that simulated drill rows resulted in better establishment than broadcasting. In the northern part of the Sultanate it has been demonstrated that Rhodes grass could be established and a rotational grazing system using sheep and goats applied in a model simulating traditional farms. Trials on irrigated pastures and fodder crops were conducted in various sites in the country by the Department of Agricultural Research and by IRI. These trials include tropical grasses and legumes, alfalfa, maize, hybrid napier Sudan grass and multi-cut sorghum. Some adaptable and highly productive species and varieties were identified.

In the United Arab Emirates agronomic trials were carried out using Rhodes grass, blue panic, elephant grass and barley.

The arid rangelands of the GCC member countries have several common characteristics and pose similar challenge to proper use and conservation. Although national programmes are being developed, yet there is an urgent need to establish or strengthen joint action aiming at stimulating collaborative efforts to conserve and improve rangelands, and increase fodder crops and pasture production.

2. OBJECTIVES

(a) Long term objectives
A joint programme will be created in which all countries of the region will actively participate in order to arrest rangeland deterioration and promote regeneration of range vegetation for the purpose of soil protection, forage production and other uses. Ultimately the red meat- and milk-supplying power of the region's arid rangelands will be increased for the benefit of rural and urban communities.

(b) Immediate objectives
(i) Facilitate the exchange of information, material and practical experience gained in the fields of range and pasture management and fodder crop production between member countries of the GCC. (ii) Elaborate rangeland management and improvement methodology based on experiences of member countries but which can draw whenever necessary from adaptable practices and appropriate technologies from other regions. (iii) Identify and establish pilot areas for demonstrating integrated development of rangelands with the active participation of bedouins, farmers and livestock owners. (iv) Strengthen current training and extension programmes with the objectives of increasing the availability of well trained national professionals and technicians. This will be coupled with supporting activities to educate land users so that conservation and improvement are achieved and maintenance of a productive range resource is ensured. (v) Identify means and adopt techniques for overcoming constraints, and improve feed availability from irrigated fodder crops, pastures and adaptable fodder trees and shrub to supplement natural forage resources.

3. PROPOSALS

3.1 National institutions
There is a necessity to create well-defined units within corresponding ministries or to strengthen present ones dealing with range, pasture and fodder crops in members countries. These units should be equipped with qualified staff and given authority, funds and responsibility of planning and implementing different programmes in the fields of range and pasture management and fodder production. These units should undertake actions to protect range resources and preserve rights of use by various beneficiaries without its degradation. Attention should be given to the creation of a well-equipped range, pasture and fodder crop extension quadre to assist bedouins, farmers and private enterprise in developing properly managed and productive systems. This extension quadre will be responsible for promoting the required awareness among livestock owners in order to conserve and properly use the grazing resources.

3.2 Conservation and utilization of germplasm of native forage species
The region is endowed by the presence of several annual and perennial forage species which are adapted to aridity and/or salinity (Appendix 1).

Many of these species are becoming less abundant due to overgrazing and vegetation clearance as a result of human activities. Germplasm material from these species is of great value to member countries and has a good potential for utilization in other regions. This germplasm resource should be collected, described, catalogued and preserved as a preliminary phase. In a following phase investigations should be conducted to determine practical ways of using suitable material for range improvement, pasture establishment, browse, sand dune fixation, shelterbelts establishment, etc. A computerized system should be developed to store all information and data about the collected and studied germplasm to facilitate accessibility to member countries.

3.3 Watershed management

The importance of rangelands does not solely rest with the forage available for livestock since the type of natural vegetation which occurs and the kind of management it receives contribute greatly to preventing soil erosion and to the efficient harvesting of precipitation. It follows from this that some areas should be managed as watershed where the first priority will be given to maintenance of vegetative cover for controlling soil erosion, enhancing the recharge of underground water supplies and controlling the uniformity of surface flow. Several examples can be cited for common watershed and underground or surface water flows that occur between GCC member countries. Vegetation cover degradation by grazing or felling these important areas may virtually mean loss of valuable productive soil and retarding underground water recharge which translates into a reduction of the already scarce water resource. This can adversely affect water users in the same and in neighbouring countries. Arising from this fact and considering the universal agreement that in GCC member countries water is the most important single constraint of agricultural development, watershed areas should be delineated, and coordinated programmes for their management and conservation promptly started.

3.4 Fodder trees and shrubs

There is an important role being played by browse species in the provision of a good proportion of animal feed in the region mainly to goats and camels, but also to sheep. Species of *Acacia tortilis*, *A. ehrenbergiana*, *Prosopis cineraria*, *Maerua crassifolia* and *Anogeissus* are subjected to great pressure from grazing animals as well as clearance to put land under irrigated agriculture. Conservation, assisted regeneration of these species and planting them in former habitats need to be given immediate attention. This is initially interpreted as efforts to collect seed, establish nurseries where seedlings are produced from both seed and cuttings, and investigate efficient methods of propagation and establishment in the field. Ultimately these activities could be directed to large scale utilization of these species in shelterbelts, sand movement control, range rehabilitation and creation of fodder reserves. Drawing the attention to the usefulness of indigenous fodder trees and shrubs should, by no means, underestimate or discourage the utilization of exotic species that will suit local environment.

Recently, the introduction of fodder shrubs has been undertaken to utilize salt-affected lands and rehabilitate degraded arid rangelands. Further research and demonstration should take the following into consideration: (i) Use of optimum spacing that will encourage growth of native grasses and forbs and moderates drought effects on planted shrubs. (ii) Direct seeding should be applied whenever environmental conditions and site characteristics allow. Work is required to identify optimum conditions and practices for the success of directly seeded shrubs. Important among these conditions and practices are the amount and distribution of rainfall and the level of soil moisture sufficient to trigger germination and assist early establishment. Determination of optimum conditions for direct seeding will result in a rapid expansion of areas planted with fodder shrubs and consequently a significant increase of forage supply at a reasonable cost. (iii) Soil moisture storage could be increased by applying adaptable techniques to enrich the soil profile from the sporadic rainfall. (iv) Persistence of and productivity from etablished fodder shrubs will always depend upon adoption of efficient management/utilization plans that will optimize forage yield and livestock production without any deleterious effects on the stand. (v) Fodder shrubs plantations may offer a more variable forage which can best be utilized by an animal mix.

3.5 Pasture establishment and management

Direct grazing of a grass/legume pasture with low water requirement can considerably reduce the cost of production of animal products from traditional and small farms as both labour and water are possibly the most expensive inputs. Additional trials on compatability of the already proven grass and legume species, cultural and management practices should commence immediately. Mixtures should always be compared with stands of pure alfalfa and N-fertilized perennial grass. Legumes may include species such as *Clitoria ternatea*, *Macroptillium atropurpureum*, *Leucaena leucocephala*, *Medicago sativa* and *Trifolium fragiferum*, while grasses such as *Chloris gayana*, *Cynodon dactylon*, *Panicum maximum*, *P. antidotale*, *P. coloratum*, *Cenchrus ciliaris* and *Festuca arundinacea* are recommended. Inoculation of legumes with the specific *Rhizobium* strain is required to establish well nodulated legume component in the pasture system. Economic evaluation of directly grazed pastures *v*. the traditional cut-and-carry system of fodder production should be carried out taking into consideration the different forms of subsidies prevailing.

3.6 Fodder crops

Information available in the region about adaptable and productive grasses, legumes and cereal forages and appropriate agronomic techniques should be used to demonstrate to local farmers the value of these fodder crops and the role they can play in increasing livestock feed. Also, it should be demonstrated that by using some of these crops a more uniform forage supply for farm animals and local markets could be achieved throughout the year.

A useful practice which should also be developed and demonstrated is the possibility of undersowing date palms with fodder and pasture species with favourable effects on production of dates and forage. This practice could be extended to areas under fruit trees. Additional trials may be required to determine optimum cultural and management practices under specific site conditions.

3.7 Alfalfa production and management

Realizing that alfalfa is the most important fodder crop in the region and that several local types (Hejazi, Hesawi, Bahraini, Quriyati, Suri, Batinah, Basri) have been cultivated for centuries in this area, it will be natural to direct part of the research activities towards germplasm characterization, optimum agronomic practices, tolerance to stress factors, resistance to pests and diseases so that an improvement is achieved for both forage and seed production. The following actions are required to be taken: (i) Collection, cataloguing and characterization of germplasm of locally available material in order to maintain genetic purity and conduct controlled crossing. (ii) Varieties which are high yielders elsewhere should be subjected to adaptive research prior to planting on large scale. (iii) Demonstration of possibility to obtain high yields by using relatively low seed rates. Traditionally, seed rates are very high probably as a result of low seed quality available in local markets and the common practice of broadcast sowing of alfalfa which is liable to waste a large amount of seed. (iv) Proper irrigation management, i.e. amount, frequency of application and water quality. (v) Nodulation and inoculation technique studies particularly in newly reclaimed or highly saline soils. There are indications of poor nodulation and use of soil from old alfalfa fields for inoculation. (vi) Harvesting management, i.e. height, frequency and method of cutting. Some local alfalfa types are not adaptable to mechanical cutting at 5–7 cm above soil level as the crown buds are located directly below the soil surface and are thus adapted to traditional cutting at 0 cm level. (vii) A regional survey of disease and pest problems of alfalfa in order to recommend control measures (seed and forage are moved freely within the region). (viii) Breeding in order to transfer desirable traits to local types.

Thought should be given to alfalfa dehydration to produce pellets which are easy to handle, store and transport with minimum loss of feeding value. Pellets can then be used efficiently for livestock supplementation and poultry feeding. Alfalfa cubes are also useful to produce as a nutrition feed for the developing dairy industry in the region. Besides its outlined advantages, alfalfa processing will furnish the private sector with a promising industry utilizing locally produced material and assist in providing employment in rural areas.

3.8 Salt-tolerant fodder species

Feed deficits presently and in the future can be moderated by growing salt-tolerant grass, forb, tree and shrub species on affected soils or on sandy soils using brackish water. It has already been demonstrated in Saudi Arabia and Kuwait that salt-tolerant grasses such as *Chloris gayana* and *Panicum antidotale* grow well under irrigation with brackish water. Another perennial salt-tolerant grass, *Diplachne fusca* has grown well on saline soil in the Punjab Plain in Pakistan and may do well under similar conditions in the Gulf States. Related to this activity is the possibility of conducting trials on the local coastal species, *Avicennia marina* (mangrove) to determine the adaptations, productivity and palatability of the ecotypes present in the region and their role in supplying browse. Proper utilization of saline land and salt-tolerant fodder species can moderate feed deficits in the region.

Additional species which could be included in regional trials and demonstrations, e.g. *Puccinellia distans* (Pakistan), *Sporobolus fevilosum* (India), *Sporobolus arabicus* (local), *Distichlis spicata* (USA), *Macroptillium atropurpureum* (Australia), *Trifolium fragiferum* (USA), *Kochia prostrata, K. indica* (Egypt), *K. scoparia* (USA), *Atriplex* spp (ACSAD), *Sesbania aegyptica (Pakistan, PDR Yemen), Prosopis cineraria* (local), *P. tamarugo* (Chile), and *P. strombolifera* (Argentina).

3.9 Sewage effluent

The scarcity of irrigation water in the region and the need to recycle municipal water and utilize it for agricultural purposes without environmental hazards is referred to in several reports. Some member countries, e.g. Saudi Arabia, Kuwait, Bahrain and Qatar have already started some activities to use sewage effluent for fodder production shelter belts establishent, etc. With the emergence of new townships equipped with modern sewage systems, there should be an expanded opportunity for increasing the areas under fodder crops, irrigated pastures and fodder trees and shrubs. Experience and information gained through research or projects should be utilized on regional basis. Trials should be conducted to assist in choice of suitable local and introduced species.

Related to this issue, consideration should be given to the utilization of sewage effluent to establish *Prosopis* spp. for the production of pods (11% protein) that can be used directly as animal feed or as an ingredient for the preparation of livestock rations. The native species, *P. cineraria* and exotic species such as *P. tamarugo*, *P. chilensis* and *P. juliflora* are suitable candidates. Two of these namely *P. cineraria* and *P. tamarugo* are also salt-tolerant.

3.10 Seed production of fodder and pasture species

Seed production is mostly carried out in traditional farming areas without sufficient technical assistance and guidance to use more efficient practices in order to produce good quality seed. Seed of crops such as alfalfa, cowpea, sorghum, millet and barley is produced locally in small amounts to be used in the same farm and any surplus is sold at neighbouring markets. There is a large volume of seed importation to meet the expanding need of the region. As environmental conditions are favourable for the production of seed of several fodder and pasture species, a collaborative programme should be started for the production of seed from alfalfa, sorghum, Sudan grass, barley, millet, Rhodes grass and siratro as a first phase, to be followed by a second phase that includes other species according to local requirements. When production of alfalfa seed is undertaken, sufficient effort should be made to control pests particularly the green mirid that destroys a large percentage of flower buds, flowers and developing seed. Attention should also be given to encouraging pollinators such as the honey bee.

3.11 Biological nitrogen fixation in support of fodder and pasture legumes establishment and production

A regional programme is required to assist national institutions to acquire practical biological nitrogen fixation technology and small-scale inoculant production to meet requirements of forage legumes. Activities should comprise: (i) screening native and exotic strains of rhizobia for effectiveness and tolerance to stress factors, e.g. salinity, drought and high soil temperature; and (ii) inoculation techniques and types of inocula suitable to local conditions.

3.12 Livestock

Although oil dominates the economy of the region, yet traditional raising of livestock by nomads and settled farmers can play an important role in diversifying the economy. The significance of the livestock sector arises from conversion of native forage from arid rangelands into products consumed by both rural and urban societies, monetary returns to the owners of herds and flocks, and employment. Moreover, livestock raising has its social implications. Therefore, there is a pressing need to assist traditional livestock raising to gradually shift from subsistence to income-generating activity. The following points need to be considered on regional basis: (i) In their meeting in Riyadh, in February 1986, the GCC Ministers of Agriculture recommended conservation and improvement of local animal breeds. There is a large influx of goats, sheep and cattle into the region for slaughter and care must be taken to avoid uncontrolled crossing with local types of animals which are adapted to the environment. (ii) Provision of credit facilities to traditional livestock owners. (iii) Availability of well defined marketing systems that provide sufficient incentives to pastoralists and livestock owners. (iv) Maximum utilization of agricultural and industrial by-products in animal feeding. (v) Restructuring of herds and flocks to contain fewer non-productive animals and to dispose of any surplus. (vi) Strategic supplementation of range-livestock to ensure adequate nutrition during critical periods.

4. RECOMMENDED PROGRAMME AND ORGANIZATION

Reference is made to the organizational plan presented in Fig. 1.

Agricultural Sector

The whole organization comes under the Agricultural Sector of the GCC Secretariat.

Review Committee

A broad membership committee that meets biennially to review the entire programme, decides on required changes and approved budgets. This

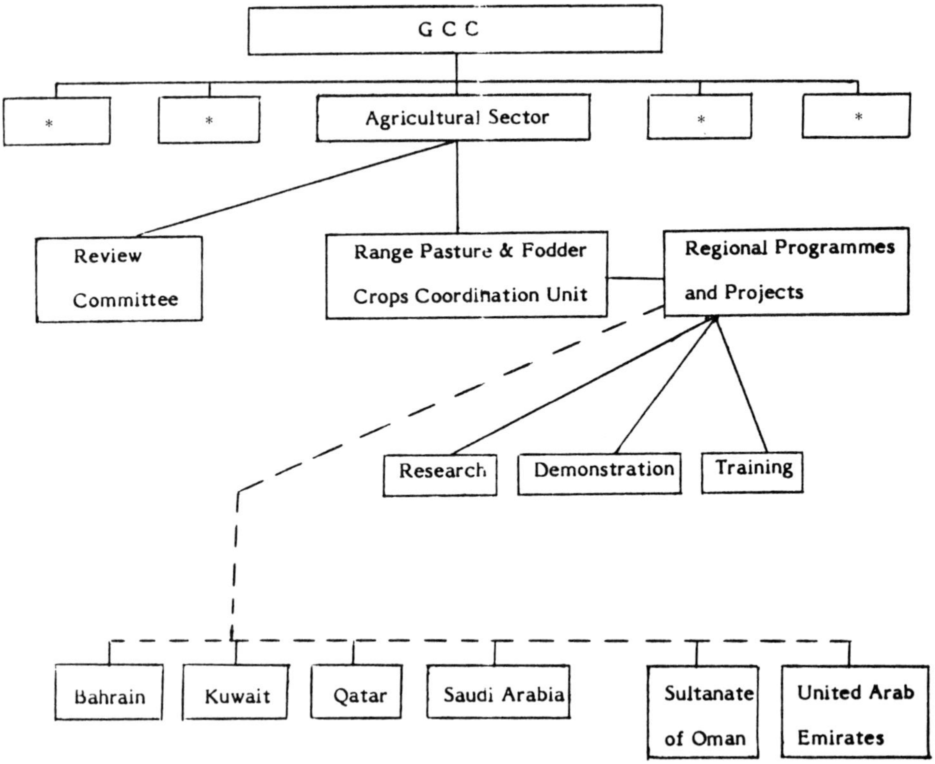

Fig. 1. Proposed organization for the regional range management programme under the Agricultural Sector of GCC.

* Other sectors which do not concern the present study.

committee can include representatives from member countries: agricultural sector, head of the range coordination units, project managers or team leaders and private sector as well as FAO Range, Pasture and Fodder Crops Coordination Unit. The Coordination Unit should be stationed with the Agricultural Sector and should perform follow-up and catalytic functions for the programme activities. The Head and other specialists of the Unit should pay regular visits to institutes, research stations and projects within member counties in regard to the above activities. The Unit is expected to facilitate the exchange of seed material and propagules between different researchers and projects staff in the member countries and between them and organizations outside the region. It should also organize the publication of a bulletin containing technical reports concerning the research and demonstration activities. The Head of the Unit is also responsible for the preparation of an annual progress report and in collaboration with an evaluation team prepares a biennial report to be presented at the Review Committee meeting.

Evaluation Team

A multi-disciplinary range ecology, socio-economics, animal husbandry, forestry, pasture and fodder agronomy consultants team is required to tour the member countries, evaluate the on-going activities and prepare, with the Head of the Coordination Unit, a biennial report to be presented at the Review Committee meeting.

Projects
Various projects in the fields of range and livestock development, revegetation and reversing land degradation are required.

4.1 Research
This comprises basic studies of range resources, range utilization and grazing systems, range improvement through agronomic practices and soil moisture conservation techniques, livestock management, use of local feeds and agricultural by-products for supplementary feeding, fodder crops, fodder shrubs and trees, irrigated pastures and wildlife in rangelands.

Considering the coordination between the member countries of the region in various fields and in order to pursue the declared goals of utilizing the resources of the region to the benefit of member countries, existing research facilities as well as those under construction and near completion should be organized in a manner that covers the research needs of similar environments and ecological zones of the region. To increase the efficiency of the research work conducted within a regional context, the present institutions can be divided into regional centres and satellite research stations.

(i) *Regional centres:* these include both Al-Jouf-Sakaka Range and Animal Development Research Centre, and Kuwait Institute for Scientific Research. These centres should be reinforced to form the major institutions for conducting research in range, pasture, fodder crops and range-livestock systems.
(ii) *Satellite research stations:* they include the Agriculture and Livestock Research Stations in Rumais and Salalah (Oman), Abu Samra Sheep Farm (Qatar) and the Regional Research Institute for Agriculture and Water Resources (Saudi Arabia). Depending upon facilities and environmental conditions, these satellite stations could be responsible for specific research programmes according to national and regional priorities.

Considering the importance of the desertifying effects of vegetation removal, excessive grazing, salinization and overpumping of groundwater aquifers, a new Revegetation and Desertification Control Centre is proposed to be established near Al-Ain, U.A.E., to lead the research and demonstration activities in this important area. Indeed some member countries are channelling funds and putting good efforts into combating desertification. In some areas, activities on proper use and improvement of rangeland coupled with the control of sand movement are being initiated. Hence the Centre can also play a coordinating role to facilitate exchange of information and experience between the member countries. The Centre will emphasize practical approaches to the rational utilization of range vegetation and appropriate techniques of range improvement.

Collaboration of these centres and stations with the various universities and institutes in the region should be encouraged particularly in relation to basic research and M.Sc. programmes and their associated experimental projects.

4.2 Demonstration

The purpose of demonstration is to illustrate to local groups the benefits from proper range use, rangeland improvement, fodder crop production, better hygiene and management of livestock. In range areas of importance to bedouins and livestock owners, suitable sites will be identified for the establishment of pilot demonstration units where practical range conservation and improvement measures can be applied with the involvement and active participation of livestock owners. As much as possible, use should be made of traditional range use and decision-making with regard to grazing and livestock movement. It has been realized that part of the difficulties faced by the various attempts of range seeding in arid lands stems from the fact that livestock owners are not actively involved in this programme, thus the reseeded areas were not given adequate protection. There is no doubt that the best type of range protection is that offered by the range users themselves. In implementing this demonstration programme, consideration should be given to the following: (i) The local practice of range *hema* could offer a suitable and acceptable system for the introduction of range conservation and improvement. *Hema* is an Arabic word meaning protection. It is applied both to a system of protection and management of grazing lands, and also a specific area reserved and managed according to that system. (ii) During the initial period of implementing rangeland rehabilitation and grazing management plans, supplementary feed is required to sustain livestock while efforts to improve the range or allow early growth and establishment of range species are in progress. (iii) As these arid rangelands at their present stocking rate cannot be expected to provide year-long livestock feed, it will be necessary to identify within selected sites, areas where fodder crops and/or fodder trees and shrubs can be grown in order to provide supplementary forage to meet the feed deficits on the range. Runoff harvesting, supplementary irrigation and soil moisture conservation techniques can be utilized to produce this additional animal feed. Thoughts could be given to combining a demonstration site with an irrigated agriculture area which will act as a source of supplementary feed and fattening finishing ground that can absorb surplus animals and prepare them for the market.

4.3 Training

Lack of qualified range management national staff, both professionals and technicians, is an important obstacle towards implementing sound programmes of range resource management in an expanded manner. The region has a good nucleus of high education institutions that can be employed to perform an efficient role towards supplying the required staff. Local as well as external training can also be periodically carried out to upgrade the technical know-how of the national staff.

(a) At the B.Sc. agricultural level: at least one university is required to establish a department of range science, agronomy and range management or forestry and range science where some students can acquire B.Sc. majoring in range science. Supporting courses of plant ecology, taxonomy, aridland forestry and range livestock management can be offered concurrently with range sciences. The same university can gradually initiate M.Sc. programmes in the fields of range sciences, pasture management and fodder crops production in conjunction with on-going projects and research stations. One or two universities of international reputation in these fields can be contacted to assist in such programmes. It may be advisable to direct M.Sc. programmes to solve problems of local nature.

(b) Diploma-certificate level: at the intermediate level, there are two institutions that can be assisted to prepare technicians to work in the fields of range management, pastures and fodder crops namely, the Veterinary and Animal Production Training Centre in Al-Hofuf (Saudi Arabia) and the Nizwa Agricultural Institute (Oman). A prerequisite to achieve this goal is to reinforce the range, pastures and fodder crops programmes in these institutes. They can also play an important role in short-term training in the above-mentioned areas.

(c) In-service training: staff joining range departments, range research institutions and projects should be given some orientation and on-job training by specialists and experts attached to the different institutions.

(d) Short-term training: such training courses should be problem analysis-oriented with sufficient practicals and field discussions and minimum lecturing. The duration of such courses could range from 4–8 weeks, inviting participants from member countries. It is expected that both local and visiting specialists collaborate in conducting the training.

(e) Study tours: professional and technical staff should be involved in such tours and visits of short (4 weeks) duration outside the region to relevant projects and international research organizations.

(f) Bedouins and other land-users training: improved production from arid rangelands and the different types of livestock utilizing them require assisting these groups to acquire additional skiils and better understanding of the potential of the land areas they occupy. Hence it is important to conduct demonstrations and short training sessions involving pastoralists and farmers to achieve this improvement. Training is expected to be in the form of simple information and practical activities that demonstrate useful methodology of hay making, management of fodder crops, planting fodder trees and shrubs, rural handcrafts and performing simple veterinary treatments.

LITERATURE CITED

FAO. 1983. *FAO Production Yearbook,* vol. 37. FAO, Rome, Italy.

GRM International Livestock Inventory. 1982. Range and livestock survey. Ministry of Agriculture and Fisheries, Muscat, Sultanate of Oman.

Kernick, M. D. 1966. Plant resources, range ecology and fodder plant introduction. Report to the Government of Kuwait. FAO, Rome, Italy.

International Research Institute. 1975. Annual report. Ministry of Agriculture and Fisheries, Muscat, Sultanate of Oman.

Taha, F. K., M. A. Razzaque, A. R. Suleiman and A. A. Nassef. 1984. An integrated approach for animal production from range ecosystems: The Kuwait experience. Second International Rangeland Congress, Adelaide, Australia.

Appendix. Native forage species*.

(A) Grasses

Apluda mutica	A
Avena barbata	A
Bothriochloa bladhii	A
Brachiaria eruciformis	A
Bromus spp.	
Cenchrus ciliaris	P
Cenchrus pennisetiformis	P
Cenchrus setigerus	P
Chloris virgata	A
Chrysopogon aucheri	P
Cymbopogon schoenanthus	P
Cynodon dactylon	P
Dactyloctenium aegyptium	A
Dactyloctenium scindicum	A
Dicanthium annulatum	P
Dicanthium foveolatum	A
Digitaria ciliaris	A
Eleusine compressa	P
Elyonurus hirsutus	P
Eragrostis barrelieri	P
Eragrostis cilianensis	A
Eragrostis spp.	
Eustachys paspaloides	A
Hyparrhenia hirta	P
Lasiurus hirsutus	P
Loudetia flavida	A
Panicum coloratum	P
Panicum turgidum	P
Pennisetum divisum	P
Pennisetum setaceum	P
Poa sinaica	P
Setaria pumila	A
Sporobolus arabicus	P
Sporobolus spicatus	P
Stipa capensis	A
Stipagrostis ciliata	P
Stipagrostis plumosa	P
Themeda quadrivalvis	A
Urochloa panicoides	A

(B) Sedges

Cyperus conglomeratus	P

(C) Forbs and herbaceous species

Alysicarpus vaginalis	P
Artemisia herba-alba	P
Lotus pusillus	A
Medicago aschersoniana	A
Medicago spp.	
Onobrychis ptolemaica	P
Plantago boissieri	A
Rhynchosia minima	*P*
Tephrosia humilis	P
Teramnus repens var. *gracilis*	P
Tribulus macropterus	A
Trigonella anguina	A
Vigna radiata	A

(D) Trees and shrubs

Acacia ehrenbergiana
Acacia tortilis
Acacia tortilis sub-species *raddiana*
Atriplex halimus
Atriplex leucoclada
Calligonum canescens
Capparis decidua
Gymnocarpus decander
Haloxylon persicum
Haloxylon salicornicum
Indigofera spinosa
Maerua crassifolia
Omocarpus dhofarica
Prosopis cineraria
Retama raetam
Rhanterium epapposum
Salsola tetrandra
Salsola vermiculata
Salvadora persica
Ziziphus spina-christi

*There are several other species that can be added to this list.
A = Annual.
P = Perennial.

Water harvesting and management in arid lands

J. L. DODD AND Q. D. SKINNER

Department of Range Management, University of Wyoming, Laramie, Wyoming 82070, USA

ABSTRACT

The purpose of this paper is to review the major ideas associated with harvesting and managing water for enhanced primary productivity on arid rangelands. Since productivity of arid rangelands is generally most limited by plant-available water supplies, the main opportunities for improvement in productivity come from more effective management, conservation and use of the limited amount of precipitation that occurs in an area. Our paper discusses the major theories and concepts pertaining to collection, movement and storage of water in rangeland substrates and to the use of water by plants in arid rangeland ecosystems. We also discuss the mechanisms and relative effectiveness of various range improvement practices that have been developed to enhance productivity via manipulations of the water cycle. Finally, we briefly discuss implementation of water harvesting practices on rangelands of the Arabian Peninsula.

INTRODUCTION

Range improvement practices dealing with water harvesting and conservation must be adaptable to local biological, physical and socio-economical conditions. In this paper we review ideas and principles of water harvesting and management that can lead to enhanced primary productivity. We make some tentative suggestions as to the applicability and feasibility of implementation of practices based on these principles in the rangelands of the Arabian Peninsula. Final decisions on application of these principles must come from specialists that are familiar with both the human and the range resources of the area. Range improvement and development of most types fail when local social and cultural traits are not carefully considered (Fortman 1985, Taha 1985, Sandford 1983).

Our use of the term water harvesting differs from the traditional usage (Heady 1975, Stoddard *et al.* 1975, Branson *et al.* 1981). It has generally been used to refer to collection of water from artificial or natural watersheds and storage in rain traps, guzzlers, livestock tanks or reservoirs for use by livestock or native fauna. We use the term to also include water collection from large (several km^2) or small (as small as a few cm^2) watershed areas with storage in the soil profile for plant use. The latter includes the harvesting that traps and concentrates water in depressions created by processes ranging from hoof action to mechanical treatments such as pitting and land imprinting (Branson *et al.* 1981, Pratt 1983). Range improvement practices to harvest water for enhanced productivity in arid regions of the world should be designed with an understanding of four hydrologic processes; channelization, infiltration, depression storage, and losses of stored water.

DESCRIPTION OF HYDROLOGIC PROCESSES

Channelization

Channelization refers to the formation of open conduits (channels) by concentrations of water during overland flow events. It can be detrimental or beneficial to harvesting water. It is detrimental when channels move water from a local area where water storage in the soil profile is desired and beneficial when one wants to move water from a watershed to an area of concentration. A basic understanding of the role channels play in transport of water can be explained using the "variable source" area concept described by Hewlett and Nutter (1970) and expounded on by Branson *et al.* (1981). This concept indicates that initial gains or losses to stream flow are from or to banks of channels. Recharge or discharge to or from channel interspaces or areas distant to channels will be a delayed response. Flow from areas with many channels clustered together should respond to precipitation events quicker than in areas with few channels. Many channels tributary to each other should collect water faster because of less interspace and more channel area. Flow from channels on steeper slopes will move downstream faster than on flatter terrain. More tributary channels will develop on steeper than on flatter terrain (Schuman 1956). In arid regions of the world, stream flow is often ephemeral (flow events occur only in response to local precipitation events) and, if the area is channelized, water moves through surrounding terrain quickly resulting in a potential loss of soil water storage in the space between large contributing channels. This is especially true if small contributing channels are developed enough to cause increased drainage of overland flow from this interspace. In this case channel bank areas may be the only area available for harvesting water. In contrast, if range management surface modifications practices are designed to eliminate small channel flow from the interspace between larger channels, water may be harvested locally. If not, then most local water storage will be concentrated only along larger channels because of channel transmission loss of flow.

In arid regions, channel transmission loss of water often occurs to the immediate area surrounding the channel during any one runoff event (Lane *et al.* 1970). This phenomenon is important because water is stored locally in the channel bank. Storage capacity is related to type of substrate material surrounding the channel, length and sinuosity of the channel, and channel maturity. Glymph and Holton (1969) show that in arid regions, loss of stream flow from any single runoff event should be maximum near the mouth of a drainage basin and increase with basin size when channel transmission loss occurs. Recharge to storage, however, will last only as long as flow persists in channels.

Amount of colluvium (soil material deposited from sheet flow along slopes of valleys) and alluvium (soil material deposited from flow in valley bottoms) is related to valley maturity. Channel maturity increases as valley maturity increases. Immature valleys are V-shaped, steep, narrow, and colluvium often is the dominant material along channel banks. Mature valleys are flat, wide, and terraced with floodplains. Alluvium is the dominant material surrounding the channel. Immature channels are U-shaped and tend to be straight. Mature channels are terraced and tend to meander. Meanders increase sinuosity, thus stream length per unit valley length and travel time of flow is increased (Leopold *et al.* 1964). Increased travel time should maximize channel transmission loss and increase local storage of groundwater. Storage potential will depend on extent of colluvial and alluvial fill over impermeable bed substrates.

Transmission losses of water in channels cause a decrease in water flow. A decrease in flow causes a decrease in the water's power to transport suspended sediment and bed load. Consequently, in arid regions channels cut and fill in part because of climatic cycles and perhaps because of grazing management practices (Thornthwaite *et al.* 1942). The semiarid cycle of erosion (Branson *et al.* 1981) shows that in arid regions during dry periods sediment is deposited in low-gradient stream reaches. This bedload may decrease channel gradient upstream if not transferred downstream during subsequent runoff events. Aggradation of bed material will cause channel filling locally as well as upstream. Aggradation of bed material is caused in part by channel transmission loss of flow during dry periods. During wet periods increased flow and velocity degrades channels especially when bed material is not stabilized by vegetation.

Channel aggradation of sediment in wide channels causes a decrease in velocity of flow events. This reduced velocity will cause sediment to be deposited bank first during high flow events. As flow decreases, sediment will then be deposited within the channel causing increased aggradation of bed material (Andrews 1982). If vegetation stabilizes the new deposits, the channel narrows. With narrower channels less flow is needed to cause overbank flooding of floodplains (Graf 1980) and more water should be stored locally as groundwater. The geomorphology of the stream channel will be controlled by established vegetation (Leopold *et al.* 1964, Graf 1978, Heede 1981). In areas such as uplands, where flow is not concentrated in channels, erosional patterns influence where water harvesting practices should be located to increase primary production and retard channelization.

Rate and extent of erosion will determine effectiveness and longevity of improvement practices designed to harvest water. Schuman (1956) discusses zones of erosion as related to origin of drainage basins. Three zones are described: (1) the belt of no erosion, (2) the zone of active erosion, and (3) the zone of deposition. Extent of these zones will vary depending on slope, soil, and vegetation. Slope, soils, and vegetation create a stable condition within the belt of no erosion. Overland flow, if present, does not move with enough velocity to cause soil movement nor channel formation. When slope increases, vegetation decreases, or soils become less resistant to overland flow velocity, soil particles move and channels eventually form creating a zone of active erosion. As channel slope gradient decreases, vegetation increases, or soils absorb water, sediment is deposited forming the zone of deposition.

The zone of active erosion is recognized by the presence of channels or other evidence that channelization is occurring. This erosion process proceeds by first causing formation of litter, soil, and rock debris dams to pile between clumps of vegetation. Eventually as channelized water flows overland, debris dams give way to irregular patterns of pedestalled plants in a matrix bare ground. Rills may eventually form and crossbraid into discontinuous then continuous gullies. Gullies may then erode to become stream channels. All increase overland flow velocity and thus decrease local storage of water. Many surface modification practices have been initiated to stop channelization on zones of active erosion by curbing overland flow using depression storage and infiltration theory (Dixon 1975).

Depression storage

Surface modification designed to create depression storage is accomplished by creating surface roughness. Surface roughness breaks down large potential zones of active erosion into many depressions on microwatersheds, each of which has its own isolated erosion zones draining to a small closed basin which impounds water. Rough surfaces have more air/soil interface area than smooth surfaces. This in turn, increases potential infiltration rates.

Infiltration

Infiltration, the process of water crossing the air/soil interface, has been incorporated into a model for predicting change in infiltration rate between smooth and rough surface profiles and degree of sealing by erosion (Dixon 1975). Dixon's model shows that infiltration is highest when: (1) surfaces are rough, (2) macropores are maximized, and (3) surface sealing is minimum. Rough surfaces increase depression storage thus increasing residence time and hydraulic head of water over soil pore spaces in one location. The presence of macropores allows water to infiltrate faster then spreads water in soil as percolation takes place to surrounding micropores. Surface sealing decreases macro and micropore openings at the air/soil interface causing a decrease of air exhaust from the soil as water fills pore space. Back pressure of trapped air decreases infiltration.

Roots of plants cause macropores. Infiltration rates are less under sod than under bunchgrasses and the rate of entry of water into soil is greater under shrubs than in interspaces beteen them (Blackburn *et al.* 1982). Removing vegetation to cause depression storage may initially cause increased infiltration but as roots degrade and root macropores fill that rate may decrease. It may therefore be advisable to restrict total area occupied by depressions in order to preserve macropore space caused by roots. When depressions are created and vegetation is removed, zones of erosion from the edge of the depression to the bottom exist. Depending on the soil type, erosion of side slopes may fill the depression or create surface sealing thereby causing decreased infiltration and a potential for increased channelization.

Loss of stored water

Harvested water stored in arid regions after infiltration occurs can be lost by (1) percolation to deep aquifers, (2) evaporation, and (3) transpiration by plants. It is possible to recover water from deep aquifers but it is difficult and expensive. Water stored in shallow or deep aquifers reduces evaporative losses compared to surface storage. Transpiration loss depends on where, when, and how much water can be tapped by plant roots. Water stored in soil profiles is readily available for use by plants. Roots of plants often reach water stored in shallow aquifers but water which percolates below root zones will not be transpired. With this in mind, water harvesting practices can be initiated to enhance water storage for plant use by modifying: (1) the soil surface, (2) the channel and, (3) the drainage basin. Modifications of the soil surface will enhance local water storage in root zones and modified channels will collect surface water at some point of concentration. Basin modifications will create groundwater storage after infiltration and percolation occurs within floodplains.

PRACTICES

Surface modification

Several surface modification practices which manipulate the water cycle have been developed by range managers in the past several decades (Vallentine 1980, Heady 1975, Stoddart *et al.* 1975, Branson *et al.* 1981). Most of these practices enhance productivity or reduce erosion or accomplish both ends. The most common of these procedures are pitting, ripping, chiseling, and contour furrowing. Gouging was introduced by mined land reclamation researchers (Hodder *et al.* 1972). A more recent development has been land imprinting (Pratt 1983, Dixon and Simanton 1977). Finally, the intentional manipulation of the water cycle through regulation of the amount, timing, and spatial pattern of the hoof action of grazing animals has received significant attention in the past several years (Savory and Parsons 1980).

Surface manipulations have been studied for over 40 years on native rangelands in North America (Barnes 1950). Treatments such as ripping, pitting, chiseling and contour furrowing on native sites often have increased herbage production, particularly when combined with a seeding treatment (Branson *et al.* 1962, Tromble 1976, Simanton *et al.* 1978). The chief physical benefits of these practices are threefold: (1) reduction of surface runoff and soil erosion, (2) collection and movement of harvested water to desired target areas, (3) concentration of water in localized areas, allowing longer time periods for infiltration to occur and deeper penetration of the water (Branson *et al.* 1966, Soiseth *et al.* 1974, Wight 1976, Neff and Wight 1977). While increased infiltration rates often occur in conjunction with these practices, the effect is usually temporary (Wight 1976). These techniques can have a significant positive effect on native range production and species composition (Whitefield and Fly 1939). Impacts on aboveground production are variable, with increases ranging from 0 to 270% depending on site-climate-treatment interactions (Wight *et al.* 1978a and b, Wight and Siddoway 1972, Branson *et al.* 1962).

Surface manipulations are modifications of the soil surface into configurations which harvest water and reduce runoff by increasing surface detention and storage, thereby enhancing infiltration. Their use is aimed at minimizing erosion on slopes, trapping water, and creating favourable microsites for the establishment of selected seeded species. Hodder (1969), Hodder *et al.* (1972) and Sindelar (1978) found that a surface-depression treatment can effectively recharge barren and drastically disturbed rangelands or minelands. Dollhopf and Goering (1982) noted that runoff-induced erosion was reduced substantially by the use of dozer basins and gouges, when compared to a chisel plowed control. Dollhopf *et al.* (1982) have suggested that gouging produced more consistent and slightly better vegetation establishment than dozer basins or chisel treatments.

Most of the surface modification treatments are described in more detail in Branson *et al.* (1981) and Larson (1980). The contour furrow consists of parallel ditches constructed along a slope contour, and often a series of evenly-spaced intrafurrow dams. Furrows can potentially harvest and store large amounts of water. Pitting and gouging are series of small basins about 1 m long, 0.3 m wide and 0.2 m deep arranged in the contour. The Hodder gouger was developed in Montana (Larson 1980, Hodder *et al.* 1972) specifically for use on mined lands while pitting was developed for native rangelands. Pitting and gouging both minimize runoff and trap and concentrate water. However, they are less severe treatments and are shorter lived than contour furrows. Chisel plowing is a common cultivation practice on dryland farms. The chisel shanks fracture soils to approximately 20 cm and leave the surface relatively rough. The ability of chisel plowing to trap water is considerably less than furrows or gouges, and is shorter lived.

Land imprinting is a practice where a heavy drum containing angle iron projections on the surface is rolled across rangeland to create depressions which concentrate water and increase depression storage capacity (Pratt 1983, Dixon and Simanton 1977). The depressions created by the imprinter are 10 to 15 cm deep and are of variable geometric shapes and cross-sectional areas. This procedure has not been tested extensively but appears to have a great deal of promise on many types of rangeland. In contrast to the other mechanical procedures mentioned, imprinting requires relatively low energy expenditures in imposing the treatment (Larson 1980).

Hoof action of large ungulate herbivores also has the potential of increasing depression storage and increasing infiltration rates (Savory and Parsons 1980). However, many studies have summarized the net effects of grazing to be reduced infiltration rates due to soil compaction and litter removal (Branson *et al.* 1981). This issue needs considerable research before recommendations can be made.

Channel modification

Primary production can be increased by several channel modification methods. Heede (1978, 1982) shows how to reclaim gully networks on low-gradient gully reaches by placing the check dams as close to downstream tributaries as possible, and by removing grazing pressure to allow vegetation to stabilize sediment deposited up channel from dams. He notes that after gullies stabilize with vegetation, surface flow of water occurs where it did not before because of local storage behind dams. Water storage and sediment deposition behind dams placed in gullies in desert regions have occasionally been used to grow crops when vegetation was not present in surrounding uplands.

Larger dams have been used to store water in reservoirs and are discussed by Branson *et al.* (1981). Instead of plugging gullies, entire water courses are plugged to capture water and sediment. Reservoirs filled with sediment are colonized by vegetation and primary production is increased. When the reservoirs are not filled with sediment, surface storage can be used to irrigate downstream and increased production can occur along the water's edge.

A third type of structure commonly used to control channel flow and sediment is a water spreader. French (1978) discusses water spreading as used to increase primary production on rangelands around the world. Water spreaders divert flow from channels across valley floors to increase soil water content and promote vegetation growth.

Walter (1979) points out that harvesting water in channels in desert regions has been practiced for centuries. Natural vegetation distribution in areas of precipitation of less than 100 mm is not diffuse but is contracted, i.e. it tends to be concentrated in gullies and depressions. This distribution pattern results from groundwater storage in lowland areas caused by runoff from rare storm events occurring over desert crust. Desert crust reduces infiltration and thus increases runoff to more permeable sandy draws and depressions. According to Walter (1979) concentration of the water on a smaller area can result in more total production for the watershed than if it were diffuse.

Basin modifications

Walter's (1979) idea of concentrations of overland flow in lowland areas and the subsequent response of vegetation can be used to promote basin modifications of stream channels to harvest water. For example, in the arid United States, water harvest practices have been designed to manipulate composition of plant communities growing along streams to increase water yield by: (1) removing undesirable phreatophytic plants (Gatewood *et al.* 1950, Robinson 1965, Culler 1970), (2) removing desirable phreatophytic plants (Bowie and Kam 1968), (3) using chemical sprays to depress transpiration (Davenport *et al.* 1982), and (4) removing plants growing on slopes to recharge stream side zones and augment flow.

The large savings of transpired water from removal of undesirable phreatophytic plants that were predicted by Gatewood *et al.* (1950) and Robinson (1965) were found to be lower by Culler (1970). In the last decade, however, removal of stream side vegetation to increase water yield has seldom been attempted (Graf 1980). This perhaps is because: (1) increased yields of water are short-term and are lost when stream side plants reinvade cleared zones, (2) it is too costly (Graf 1980), (3) demand for use of stream side zones is high, and (4) downstream loss of recoverable water to deep aquifers (Davenport *et al.* 1982). Graf (1980) points out that saving water by reducing transpiration is not as popular as management of the habitat for other uses. However, Graf (1980) also points out that if stream side vegetation is left along stream channels, flood events will occur more frequently because of decreased channel size and vegetation resistance to flow. Increased flooding can repair degraded stream channels, increased storage of groundwater, and increase production for grazing and wildlife habitat.

In arid regions vegetation establishment is difficult. Replacing one plant species because it uses more water than another can succeed only after it is proven that the introduced plant can be established. If it cannot, then vegetation cover may be decreased and channelization is more likely to occur. Within drainage basins, the stream side zone offers greater opportunity to manipulate vegetation for water harvesting than others. This is because vegetation may dictate geomorphology of stream channels by stabilizing deposited sediment, causing channel narrowing, and promoting overbank flooding (Leopold *et al.* 1964, Graf 1978, Heede 1981, Andrews 1982). Contrary to past thinking (eliminating plants that use more water than others within stream side zones) existing plants should be managed to create flooding and thus store water in alluvial aquifers. Transpired water will be lost from storage but flooding will occur with lower peak flow events and frequency of harvesting water will increase. Overall stability of drainage basins can be obtained as Walter (1979) alludes by using natural forces to harvest water in desert regions by concentating flow in channel areas.

IMPLEMENTATION

The Arabian Peninsula rangelands encompass an array of physical conditions that range from extremely arid, sandy, barren subtropical desert sites to very productive subirrigated marshes (Taha 1985, Taha and Krueger 1981). Use of these rangelands varies from almost exclusive use by nomadic, semi-nomadic, or even sedentary pastoralists (Cole 1975) in some instances to combinations of uses in others. In many instances it is feasible to regulate the type, timing, and amount of use while in others regulation is essentially impossible. The range practices discussed in this paper are applicable on only a relatively small subset of the combinations of physical, biological, and social conditions that occur on the Arabian Peninsula. In general, these practices and modifications of them will probably prove to be feasible only on those sites that have relatively high production potential and where it is possible to regulate use of the site prior to and following the improvement.

A significant limitation associated with the use of surface modification treatments on any rangeland is longevity of the treatment. Numerous studies on North American rangelands indicate that contour furrows retain their surface geometry and increase water detention capacity for up to 25 years (Neff 1973, Wight *et al.* 1978a and b, Branson *et al.* 1966). Dollhopf *et al.* (1982) reported that gouge and chisel microtopographic features were detectable for only about three years on loose, medium textured substrates of reclaimed mined land in the Great Plains of North America. Hiller (1984) compared contour furrowing, chisel, and gouge treatments in the Great Plains and after two years found the chiseled surfaces to be smooth, the gouge depressions to be nearly filled in and the furrows to be partially filled in but stable and still quite effective in trapping and concentrating water. Larson (1980) also reported that chisel surfaces wore down rapidly. Many studies suggest that furrowing is the most persistent of the surface modification treatments and is followed by pitting, gouging, and chisel. Much less work has been done on land imprinting longevity but we expect it to be similar to gouging and chisel plowing.

Soil texture is probably the most important single factor associated with longevity of surface treatments. Barnes (1950), Dollhopf *et al.* (1982), Branson *et al.* (1966) and others have found these types of treatments to be very short-lived on coarse textured substrates and relatively long-lived on fine or medium textured substrates. Longevity of these treatments is also shortened by lack of vegetative cover, high-intensity rainstorms, wind storms, frost heaving, runoff, and vehicular and animal traffic. Branson *et al.* (1966) concluded that medium-textured soils (loams, clay loams) are generally the most favourable textures for surface modification and based their evaluation on water storage capacity, primary productivity, and longevity of the treatments.

The positioning of water harvesting practices in relation to the three zones of erosion is important. Placement of an improvement practice on the belt of no erosion results in concentrating and storing only water that falls on that site. The improvement practice located on this zone should last longer than one placed on the zone of active erosion or zone of deposition because overland flow is minimal. Location of an improvement practice on the zone of deposition harvests water falling locally as well as that passing through the zone of active erosion. Sediment deposited in this zone, however, will shorten the life of the specific surface or channel treatment. Storage potential of this zone should be high because soils are deep compared to the other erosion zones.

Improvement practices placed on the zone of active erosion are subject to overland flow. These areas are normally steeper than the other erosion zones and care must be taken to create stability. One really has two alternatives when treating the zone of active erosion. The first is to extend the belt of no erosion and zone of deposition into the zone of active erosion by placing treatments on these two zones. Such practice usually shortens the zone of active erosion. The second alternative is to break the zone of active erosion into microzones of erosion. Terracing is an example of this type of treatment. A terrace has a belt of no erosion, a zone of active erosion (on downhill side), and a zone of deposition (on uphill side). Terracing results in many zones of no erosion and zones of deposition instead of one long zone of active erosion and stabilizes soil and stores water locally.

Another group of limitations on use of water harvesting and treatments involve availability of equipment, equipment operators, and expertise to design and build the equipment. The equipment necessary to perform these treatments tends to be extremely expensive, custom made, and almost experimental in nature (Larson 1980). The chisel plow and land imprinter are commercially available in North America but the other implements will most likely need to be constructed from raw materials or result from extensive modification of existing farm machinery. The availability of welders, heavy equipment operators, and mechanical engineers involved in oil exploration and production on the Arabian Peninsula should be a major asset in the development of equipment for range improvement. Most of the success of heavy equipment development for use on rangelands in North America originated from local land managers working closely with welders and engineers to fabricate and adapt equipment for local conditions and problems.

LITERATURE CITED

Andrews, E. D. 1982. Bank stability and channel width adjustments, East Fork River, Wyoming. *Water Resources Research* **18:** 1184–1192.

Barnes, O. K. 1950. Mechanical treatments on Wyoming rangeland. *J. Range Manage.* **3:** 198–203.

Bowie, J. E. and W. Kam. 1968. Use of water by riparian vegetation in Cottonwood Ash, Arizona. Water Supply Paper 1058. Geol. Surv., Washington, D.C.

Blackburn, W. H., R. W. Knight and M. K. Wood. 1982. Impact of grazing on watersheds: a state of knowledge. *Texas Agr. Exp. Sta.* MP1496.

Branson, F. A. , G. F. Gifford, K. G. Renard and F. R. Hadley. 1982. *Rangeland hydrology.* 2nd Ed. Soc. Range Manage., Denver, Colorado.

Branson, F. A., R. F. Miller and I. S. McQueen. 1962. Effects of contour furrowing, grazing intensity and soils on infiltration rates, soil moisture and vegetation near Ft. Peck, Montana. *J. Range Manage.* **15:** 1151–1158.

Branson, F. A., R. F. Miller and I. S. McQueen. 1966. Contour furrowing, pitting and ripping on rangelands of the western United States. *J. Range Manage.* **19:** 182–190.

Cole, D. P. 1975. *Nomads of the nomads: The Al-Murrah Bedouin of the Empty Quarter.* Aldine, Chicago.

Culler, R. C. 1970. Water conservation by removal of phreatophytes. *Amer. Geophys. Union Trans.* **51:** 684–689.

Davenport, D. C., P. E. Martin and R. M. Hagen. 1982. Evapotranspiration from riparian vegetation: water relations and irrecoverable losses for saltcedar. *J. Soil and Water Cons.* **37:** 233–236.

Dixon, R. M. 1975. Infiltration control through soil surface management. pp. 543–567. *In:* Proc., A.S.C.E. Symp. Watershed Manage. Utah State University, Logan.

Dixon, R. M. and J. R. Simanton. 1977. A land imprinter for revegetation of barren land areas through infiltration control. Proc. Amer. Water Res. Assoc. and the Arizona Acad. of Sci. Las Vegas.

Dollhopf, D. J. and J. D. Goering. 1982. Mined land erosion control with surface manipulation. Proc. Symposium on Surface Mine Hydrology, Sedimentology, and Reclamation. University of Kentucky, Lexington.

Dollhopf, D. J., D. W. Hedberg, S. A. Young, J. D. Goering, W. M. Schafer and C. J. Levine. 1982. Effects of surface manipulation on mined land reclamation. US/EPA, Cincinnati, Ohio.

Fortman, L. 1985. Range management at the grass roots: some hows and whys of local participation. p. 1–10. *In:* White, L. D. and J. A. Tiedeman (eds), Proc. Intern. Rangeland Resources Development Symposium. 13–14 February 1985, Salt Lake City, Utah, Soc. Range Manage. and Dept. For. and Range Manage., Wash. State University, Pullman.

French, N. H. 1978. Water spreading around the world. *In:* Hyder, D. N. (ed), Proc. First Rangeland Congress. Soc. Range Manage., Denver, Colorado.

Gatewood, J. S., J. W. Robinson, B. R. Colby, T. D. Hem and L. C. Halpenny. 1950. Use of water by bottom-land vegetation in lower Safford Valley, Arizona. Water Supply Paper 1103, U.S.G.S. Washington, D.C.

Glymph, L. M. and H. N. Holton. 1969. Land treatment in agriculture watershed hydrology research. p. 44–68. *In:* Effects of watershed changes on streamflow. Water Resources Symp. No. 2, Univ. Texas Press, Austin.

Graf, W. H. 1978. Fluvial adjustments to the spread of tamarisk in the Colorado plateau region. *Geol. Soc. of Amer. Bull.* **89:** 1491–1501.

Graf, W. H. 1980. Riparian management – a flood control perspective. *J. Soil and Water Cons.* **35:** 158–161.

Heady, H. 1975. *Rangeland management.* McGraw-Hill, New York.

Heede, B. H. 1978. Designing gully control systems for eroding watersheds. *Environ. Manage.* **2:** 509–522.

Heede, B. H. 1981. Dynamics of selected mountain streams in western United States of America. *Z. Geomorph. N.F.* **25:** 17–32.

Heede, B. H. 1982. Gully control: determining treatment priorities for gullies in a network. *Environ. Manage.* **6:** 4441–4451.

Hewlett, J. D. and W. L. Nutter. 1970. The varying source area of steamflow from upland basins. p. 65–83. *In:* Proc. of the Symp. on Interdisciplinary Aspects of Watershed Manage. Amer. Soc. Civil Eng. New York.

Hiller, J. G. 1984. Effects of surface manipulations on vegetation establishment on coal surface mined lands in the northern Great Plains. M.S. Thesis, Range Manage. Dept., University of Wyoming, USA.

Hodder, R. L. 1969. Colstrip reclamation project – 1969. Prog. Rpt., *Montana Agr. Expt. Sta.*, Montana State University.

Hodder, R. L., B. W. Sindelar, J. Bucholz and D. E. Ryerson. 1972. Coal mine land reclamation research, Western Energy Co., Colstrip, Montana: Prog. Rpt. 1971, *Montana Agr. Expt. Sta.* Res. Rpt. 20, Montana State University.

Lane, L. J., M. H. Diskin and K. H. Renard. 1970. Input-output relationships for an ephemeral stream channel system. *J. Hydrology* **13:** 22–40.

Larson, J. E. 1980. Revegetation equipment catalog. Vegetative Rehabilitation and Equipment Workshop. USDA Forest Service Equipment Development Center, Missoula, MT. Stock #001-001-00518-5.

Leopold, L. B., M. G. Wolman and J. P. Miller. 1964. *Fluvial processes in geomorphology.* W. H. Freeman and Co., San Francisco and London.

Neff, E. L. 1973. Water storage capacity of contour furrows in Montana. *J. Range Manage.*, **26:** 298–301.

Neff, E. L. and J. R. Wight. 1977. Overwinter soil water recharge and herbage production as influenced by contour furrowing on eastern Montana rangelands. *J. Range Manage.* **30:** 193–195.

Pratt, M. 1983. Land imprinting for lasting impressions. *Agr. Eng.* **64:** 9–11.

Robinson, T. W. 1965. Introduction, spread, and areal extent of salt-cedar (*Tamarix*) in the western states. Prof. paper 491-A. U.S.G.S., Washington, D.C.

Sandford, S. 1983. *Management of pastoral development in the third world.* John Wiley & Sons.

Savory, A. and S. D. Parsons. 1980. The savory grazing method. *Rangelands* **2:** 234–237.

Schuman, S. A. 1956. Evolution of drainage systems and slopes in badlands at Perth Amboy, New Jersey. *Bull. Geol. Soc. Amer.* **67:** 597–646.

Simanton, J. R., H. B. Osborn and K. G. Renard. 1978. Hydrologic effects of rangeland renovation. p. 331–334. *In:* D. N. Hyder, (ed.), Proc. First Rangeland Congress. Soc. Range Manage. Denver, Colorado.

Sindelar, B . W. 1978. Successional development of vegetation on surface mined lands in Montana. p. 550–556. *In:* M. K. Wali (ed.), *Ecology and the coal resource development.* Pergamon Press. New York.

Soiseth, R. J., J. R. Wight and J. K. Aase. 1974. Improvement of panspot (solonetzic) range sites by contour furrowing. *J. Range Manage.* **27:** 107–110.

Stoddart, L. A., A. D. Smith and T. W. Box. 1975. *Range management.* 3rd Ed. McGraw-Hill, New York.

Taha, F. K. 1985. Sociological aspects and range improvement: the case of Arab countries in the Middle East. p. 11–13. *In:* L. D. White and J. A. Tiedeman (eds), Proc. Int. Rangeland Resources Development Symposium. 13–14 February 1985, Salt Lake City, Utah. Soc. Range Manage. and Dept. For. and Range Manage., Wash. State University, Pullman.

Taha, F. K. and W. C. Krueger. 1981. Kuwait's range research model for development of a viable sheep industry. p. 643–665. *In:* J. T. Manassah and E. J. Briskey (eds), *Advances in food-producing systems for arid and semiarid lands.* Part A. Academic Press, New York.

Thornthwaite, W. C., C. F. S. Sharp and E. F. Dorch. 1942. Climate and accelerated erosion in the arid and semi-arid southwest, with special reference to the Polacca Wash Drainage Basin, Arizona. Tech. Bull. No. 808, USDA, Washington, D.C.

Tromble, J. M. 1976. Semiarid rangeland treatments and surface runoff. *J. Range Manage.* **29:** 251–255.

Vallentine, J. F. 1980. *Range development and improvements.* 2nd Ed. Brigham Young University Press.

Walter, H. 1979. *Vegetation of the earth.* 2nd Ed. Springer-Verlag, New York.

Whitfield, C. J. and C. Fly. 1939. Vegetational changes as a result of furrowing on pasture and range lands. *J. Amer. Soc. Agron.* **31:** 413–417.

Wight, J. R. 1976. Land surface modifications and their effects on range and forest watersheds. p. 165–174. *In:* Proc. 5th Workshop of US/Australia Rangelands Panel: Watershed Management on Range and Forest Lands. Utah State University, Logan.

Wight, J. R., E. L. Neff and R. J. Soiseth. 1978a. Vegetation response to contour furrowing. *J. Range Manage.* **31:** 97–101.

Wight, J. R. and F. H. Siddoway. 1972. Improving water use efficiency of rangelands by surface modification. *J. Soil and Water Cons.* **27:** 170–174.

Wight, J. R., F. H. Siddoway, R. J. Kertchner and J. L. Bishop. 1978b. Contour furrows seeded with grasses and legumes increase production on semiarid range sites. p. 643–644. *In:* D. N. Hyder (ed.), Proc. First Rangeland Congress. Soc. Range Manage., Denver, Colorado.

The present rangeland situation in the northern region of Saudi Arabia

ABDU AL-SHARIF

Range and Animal Development Research Centre, Sakaka, Al-Jouf, Saudi Arabia

ABSTRACT

The northern area is the biggest sheep-producing region and one of the important range areas of the Kingdom. However, its range resources have been seriously depleted. For this reason the government has paid special attention and made considerable efforts towards the development and conservation of natural resources in this area.

In the early 60's the Ministry of Agriculture started a range improvement scheme at Wadi Al-Awasi to the north of Arar by fencing an area of 6 km^2 and constructing 40 dikes for water spreading. In the same period, a range office was established at Arar. The new office was able to build several exclosures for range ecological studies, develop livestock water holes or points and assist in the distribution of free animal feed supplied to the nomadic areas affected by drought. In the late 70's and early 80's the range office built more range improvement reserves at Wadi Al-M'Leih (34 km^2), Al-Tamariat (27 km^2) and four reserves in Al-Awasi with the aim of range development and improvement by natural and artificial means of water spreading, seeding and soil conservation. The grazing reserves were intended to be open for grazing after improvement, with strict grazing monitoring.

In 1982, a range and animal development research centre was established at Sakaka, Al-Jouf. The overall objective of the research centre is the development of the range and animal resources in Saudi Arabia and the northern area in particular. This is based on research findings and an understanding of the development problems in the area. The centre has already taken substantial steps towards conducting a range inventory survey of the area and is making good progress in improvement studies at Buseita and Tamariat.

It is worth mentioning that the Kingdom has several conservation laws and regulations for the protection and development of natural resources. These regulations include the conservation of range and forest resources, as well as the protection of wildlife and the environment.

INTRODUCTION

The Kingdom of Saudi Arabia has an area of 2,253,000 km^2 and a population of about seven million people. The economy of Saudi Arabia is basically pastoral and agricultural, but it is rapidly being diversified to include manufacturing, processing and mining industries (particularly oil). Saudi Arabia (S.A.) ranks first among the world's oil producers. The oil revenues are being used for the general improvement of the country, especially all aspects of agriculture.

The rangelands are the largest natural resource of S.A. They cover about 75% of the total land area. They are vital to the national economy since they are the main grazing land for the range livestock industry. In addition, about 1.9 million nomads are dependent in one way or another upon livestock production.

The degradation of vegetation in S.A. has become a matter of great concern. Large numbers of livestock, unrestricted grazing, and wood gathering without regard for conservation have caused the destruction of native forage resources and put a severe hardship on livstock and people alike in most areas of S.A. (Kingery 1971). Yet, only recently, has the magnitude of the situation been fully realized. It is now well understood that some action to increase forage and livestock production needs to be taken. The purpose of this paper is to present the rangeland situation in the northern region of the country.

CURRENT RANGELAND SITUATION

A. Vegetation

Both man's activities and environmental interaction influence vegetation distribution (Bell 1978). The vegetation of the north of Saudi Arabia has been exposed to the harsh environment, e.g. high temperatures, low rainfall and heavy livestock use. Under these conditions it comes as no surprise that only grazing-resistant plants have remained and weaker plants have been eliminated. Most of these plants are dwarf shrubs and a few trees with deep tap roots which extract deep soil moisture. Annual grasses and herbs have a very short growing season.

Vegetation types in the north of Saudi Arabia have been studied by the Bureau de Recherches et Compagnie Nationale d'Amenagement (1983) and Heemstra and Al-Hassan (1984). The vegetation can be subdivided into six broad divisions based on soil type and topography.

1. *The Nafud:* Sand dunes of various sizes and patterns, with no defined drainage pattern. Vegetation is dominated by the shrubs *Calligonum comosum* and *Haloxylon persicum*. Other species include *Artemisia monosperma*, *Scrophularia hypericifolia*, *Cyperus conglomeratus* var. *aucheri* and *Moltkiopsis ciliata*, and the perennial grass *Stipagrostis drarii*. Among the most abundant annuals are *Eremobium aegyptiacum* var. *lineare*.

2. *The Hamad:* limestone plateau with wadis and numerous depressions. It shelters dwarf shrubs mainly of *Salsola tetrandra*, *S. chaudharyi*, *S. cyclophylla* with *Hammada salicornica*, *Astragalus spinosus*, *Achillea fragrantissima* and *Artemisia herba-alba*. In spring, annuals are abundant, including species such as *Arnebia linearifolia*, *Echinosciadium arabicum*, *Matthiola longipetala*, *Savignya parviflora* and *Astragalus* spp.
3. *Al-Harrah:* undulating volcanic massif with blocks of basalt and loamy depressions and wadis. Vegetation consists mainly of drawf shrubs dominated by *Astragalus spinosus*, *Hammada salicornica*, *Zilla spinosa*, *Achillea fragrantissima* and *Artemisia herba-alba*. In spring these are accompanied by a rich vegetation of annuals.
4. *Al-Widyan* (eastern wadis): loamy wadis between limestone hills. Vegetation is often strongly degraded and consists of dwarf shrubs including *Achillea fragrantissima*, *Artemisia herba-alba*, *Astragalus spinosus*, *Pituranthos triradiatus*, *Salsola chaudharyi*, *Hammada salicornica* and *Artemisia monosperma* (in sandy locations). Many annual forbs are present in spring in the less degraded areas. Some wadis are so strongly degraded that only the remains of dead dwarf shrubs are left and annuals are virtually absent.
5. *El-Shama:* undulating plain with a soil of shallow sand on limestone. Vegetation is heavily grazed. The dwarf shrub, *Anabasis articulata*, is characteristic of this area, accompanied by *Astragalus* spp. and *Hammada salicornica*. The annual vegetation has low diversity and consists of species which are common to most of northern Saudi Arabia.
6. *Wadi Sirhan:* a complex of alluvial soils varying from sandy to clayey, sometimes gravelly or saline. The vegetation is a complex of shrubs and dwarf shrubs. The most characteristic species of this area are *Nitraria retusa*, *Tamarix nilotica*, *Atriplex halimus*, *Frankenia hirsuta*, *Limonium axillare*, *Suaeda vera* and *S. vermiculata*. Other important range species are *Salsola cyclophylla* and *S. tetrandra*. Annuals include a few species and may be completely absent in the saline areas.

The three other important vegetation types are:

1. *Rhanterium epapposum* community. This species is dominant in a zone north of the Nafud, either on limestone plain or on shallow sand over limestone. Accompanying species include *Astragalus spinosus*, *Hammada salicornica* and *Salsola cyclophylla*.
2. *Traganum nudatum* community. This species is dominant on an undulating gravel plateau northwest of Nafud. It is accompanied by *Hammada salicornica* and in more sandy places by *Haloxylon persicum*, but often *Traganum* is the only dwarf shrub. Annuals present depend very much on soil texture and include species like *Eremobium aegyptiacum* var. *lineare*, *Mesembryanthemum forsskalii*, *Neurada procumbens*, *Arnebia* spp. and *Plantago* spp.
3. *Artemisia judaica* community. This community has very low coverage and is restricted to scattered depressions in the Buseita limestone gravel plain. Associate species include *Francoeuria crispa*, the annual *Mesem-*

bryanthemum forsskalii and other annuals which are also common in most of the country.

B. Range condition

Most of the rangelands in the northern region are in poor condition and are deteriorating rapidly. The productivity has been greatly reduced. Accurate data about forage quantity or quality are not available, which further reduces the chance of success of any development of range or livestock management. By using the regression equation for Mediterranean regions, the Bureau de Recherches et Compagnie National d'Amenagement (1983) stated that the northwest region of Saudi Arabia produces an average of 32 kg/ha consumable dry matter. This figure is based on Le Houerou (1977) for Mediteranean regions and should be adjusted before being applied to desert regions.

C. Range management problems

The north of Saudi Arabia is an arid land with a long dry season, high evaporation rate and low irregular rainfall. Large areas of rangeland can be used only during the short rainy season when the forage is good and the water supply adequate. For the greater part of the year, livestock must move to other pastures where water and forage are available. Some years are so dry that livestock have to be fed or they will perish.

Unfortunately, the management of northern Saudi Arabia's rangelands has been neglected. The continued denudation of grazing lands requires urgent attention to solve the following problems:

1. *Overstocking:*
   For hundreds of years, nomadic people have wandered from pasture to pasture in search of food and water. In the past, livestock numbers always remained small because many animals were lost through disease, periodic droughts and intertribal strife (Drabbs 1967). As a result of the new modern life and improved animal health, the numbers of livestock have increased rapidly, putting a great demand on the rangelands. The result is that there is not enough food to support current livestock numbers.
2. *Availability of transportation:*
   Because of the greater availability of trucks and tankers to move livestock and carry water, livestock owners can utilize a larger area further away from water resources. As a result of this situation, greater areas of the rangelands are made accessible.
3. *Dry farming:*
   Hundreds of hectares of arid land are used for dry farming every year. The crop yield of these farms is generally very low because soil fertility is poor. After a few years of cultivation, most of the nutrients are depleted and large areas of the rangeland become unproductive.

4. *Collection of wood for fuel:*
   Using dry grasses and shrubs for fuel is another factor causing rangeland deterioration. Wood is the principal cooking fuel in nomadic camps and a source of warmth in the cold mountains of the southwest. Because of unrestricted wood collection, most forest lands are depleted to the point that their commercial value is almost non existent (Kingery 1971).
5. *Availability of water;*
   In the past, water supply had been the limiting factor in utilization of the range, but recently the government has drilled hundreds of wells providing water for livestock in most rangelands. This encouraged livestock owners to increase their grazing areas and over-utilize the available forage.
6. *Subsidy:*
   Livestock owners can buy 50 kg of concentrate feed for US$6 and the government pays the difference in the market price. This subsidy program creates more pressure on the rangeland because subsidized grain allows animals to stay on the range for long periods without moving, thereby contributing further to the depletion of rangeland vegetation.
7. *Imbalance between animal increase and off-take:*
   It is well known that bedouins never sell female sheep, cattle, goats or camels unless they are sick or too old to reproduce. This practice results in an additional pressure on the rangelands and brings in negligible financial gain.
8. *Soil compaction*
   A great percentage of the soils in the northern area are shallow with low moisture holding capacity and limited forage production. Most of the gravel plains are cemented and crusted. A hardpan develops where crop cultivation has taken place in the depressions and drainage lines or as a result of trampling by grazing animals. Infiltration rate in these soils is very low, resulting in surface run-off and waterlogging especially in clay pockets in depression areas. As a result of soil compaction and low moisture content, large areas are devoid of vegetation.

In conclusion, it appears that droughts, combine with subsidized grain for livestock, improved livestock health, collection of wood for fuel, unrestricted grazing, greater access to larger areas, and dry farming have caused the recent destruction of the rangelands.

D. Range management activities in the northern region

The northern area is the biggest sheep-producing region and one of the most important range areas of the Kingdom. However, its range resources have been seriously depleted. For this reason the government has paid special attention and made considerable efforts to the development and conservation of natural resources in this region.

In the early 60's the Ministry of Agriculture started a range improvement scheme at Wadi Al-Awasi to the north of Arar by fencing an area of 6 km^2 and constructing 40 dikes for water spreading. In the same period, a range office was established at Arar. The new office was able to build several exclosures for range ecological studies, develop livestock water points and assist in the distribution of free animal feed supplies to the nomadic areas affected by drought. In the late 70's and early 80's the Range Section built more range improvement reserves at Wadi Al-M'Leih (34 km^2), Al-Tamariat (27 km^2) and four reserves in Al-Awasi in an attempt to improve range conditions by natural and artificial means of water spreading, seeding and soil conservation. The grazing reserves were intended to be open for grazing after complete improvement and under strict grazing controls.

In 1982, a range and animal development research centre was established at Sakaka, Al-Jouf. The overall objective of the research centre is the development of the range and animal resources in Saudi Arabia and the northern area in particular. This is based on the research findings and the understanding of the development problems in the area. The research centre has already taken good steps in conducting a range inventory survey of the area and making good progress in range improvement studies at Buseita and Tamariat.

It is also worth mentioning that the Kingdom has several conservation laws and regulations for the protection and development of natural resources. These regulations include the conservation of range and forest resources, and protection of wildlife and the environment.

LITERATURE CITED

Bell, H. M. 1978. Rangeland management for livestock production. University of Oklahoma Press, Norman, Oklahoma.

Bureau de Recherches et Compagnie Nationale d'Aménagement. 1983. Water, agriculture and soil studies of SAQ and overlying aquifers, vol. 5, 206 pp.

Drabbs, R. J. 1967. Report to the Government of Saudi Arabia on improving range production. Report No. TA2397. FAO, Rome.

Heemstra, H. and H. Al-Hassan. 1984. List of plants collected for the herbarium. Sakaka, Al-Jouf, Saudi Arabia, Range and Animal Development Research Centre, 8 pp.

Kingery, C. E. 1971. Report to Government of Saudi Arabia on the possibilities for development and management of public rangelands. A review of resources survey report supplemented with some interpretations and comments. FAO, Rome. 112 pp.

Le Houerou, H. W. 1977. Rangeland production and annual rainfall relations in the Mediterranean Basin and in the African Sahelo-Sudanian Zone. *J. Range Manage.* **30:** 181–187.

Rangelands in Oman: management, problems and prospects

TAG E. YASSIN AND SALIM S. A. ALSHANFARI

Department of Agricultural Research, Ministry of Agriculture and Fisheries, Southern Region, Salalah, Sultanate of Oman

ABSTRACT

Oman Mountains, Dhofar Mountains, wadis and desert are the main physical features which influence climatic conditions and agricultural activities in the country. Except in the mountains, rainfall is insufficient to support agriculture which depends entirely on irrigation from underground water. The main crops are date-palm, lime, alfalfa, banana, coconut and vegetables which are produced on an area of approximately 100,000 ha. There are about 125,000 heads of cattle and 70,000 camels, mainly in the Southern Region of Oman, while the number of goats is about 700,000 and that of sheep is about 135,000, both being concentrated in Northern Oman. Due to differences in climatic conditions and plant species, rangelands are divided into two main categories:

A. Northern Oman: which comprises Oman Mountains Range and its surrounding areas to the east and south. Desert and semi-desert vegetation dominates except in the mountains where other types of vegetation are found. Few studies have been undertaken, but rangelands show signs of deterioration, which can be improved by proper range management practices.

B. Southern Region of Oman (Dhofar): Dhofar, ecologically a unique green area in Arabia, has the best rangelands in Oman because of the rains brought by the seasonal southwest monsoon in June–September. Woodland, tall and short grasslands are found on the escarpments, the plateau and the foothills. Five ecological zones are recognized in Dhofar, namely: desert, semi-desert, short grassland, tall grassland and woodland, with Sudano-Deccanian element in the green parts and Saharo-Sindian element dominating the arid areas. Rangeland deterioration has taken place due to a number of factors which disturbed the Jabali peoples' pattern of land use. A programme of work is undertaken to restore and improve the rangelands.

The present paper also emphasizes the importance of research in rangelands.

INTRODUCTION

The Sultanate of Oman is the second largest country in Arabia, covering a total area of about 300,000 km^2, and has an estimated population of more than one million. The country lies in the southeastern corner of the Arabian Peninsula between 16°40′ and 26°20′ N Lat., and 51°50′ and 59°40′ E Long.

PHYSICAL FEATURES

There are three dominant physical features in Oman which have profound influence on the agricultural activities in the country. These are the mountain ranges (about 45,000 km^2), the coastal plains (approximately 9,000 km^2), and the dry wadis and the deserts (about 246,000 km^2).

THE MOUNTAIN RANGES AND COASTAL PLAINS

The Oman Mountains Range

This range, some 700 km in length, extends from Ras Masandam in the extreme north of the country in a southeastern direction down to Ras Al-Hadd in the south. Between this range of mountains and the coast of the Gulf of Oman lies the Batina Plain which is about 25 km wide. This plain is the most important agricultural area in Oman, and possibly in the whole Arabian Gulf region.

The Dhofar Mountains Range

This range extends from opposite Kuria Muria Islands in the east down to the borders with South Yemen in the west. The western part of the range, Jabal Al-Gamar, rises directly from the coast, while the central part, Jabal Al-Gara, rises sharply from the flat Salalah Plain which separates it from the Arabian Sea. In the east, Jabal Samhan rises gradually from the coast at first but then rises vertically to a 1400 m escarpment, the highest point being over 2000 m above sea level. To the north, the mountains give way to gravel hills and desert along the entire range. The top of the range is a relatively flat plateau, 700–900 m above mean sea level and about 10–15 km wide. The best rangelands in Oman are found in this range of mountains.

There are no permanent rivers, but few of the main wadis flow after periods of heavy rains, and some pools in the wadi beds remain long after the flow and sometimes throughout the year. Other permanent pools are fed by springs arising from the bottom of the wadis on the edges of the mountains. The Salalah Plain, about 40 km long and 10 km wide, is the largest and most important in the Southern Region of Oman.

THE WADIS

The northern wadis originate from the southern side of Oman Mountains and flow into the desert. Near the mountains, these wadis form important agricultural areas in which agricultural centres have developed. In some of these wadis, water flow is almost continuous throughout the year. The area encompasses some important rangelands. Further south, the wadis flow towards the Arabian Sea and Al-Wahaiba Sands in the east, or to the Empty Quarter in the south. The southern-most wadis originate from the northern side of Dhofar Mountains and flow northwards towards Najd and the surrounding desert.

THE DESERTS

For practical purposes, the whole area between Dhofar Mountains and Oman Mountains in the north, 700–800 km wide, may be considered as desert. However, there are two main deserts in Oman: Al-Wahaiba Sands in the northeast, and the famous Empty Quarter in the west.

CLIMATE

There are two climatic regions in Oman. The first region includes the Oman Mountains in the northeast of the country as well as the plains around these mountains. Temperatures are high throughout the summer and moderate in winter. On the eastern side of the mountains (Batina Plain) rainfall during winter is brought about by the northeastern winds. On the southern side of the range (Oman Interior, Dhahira and Shargia) rainfall during summer is the result of the seasonal southwestern winds. The amount of rain varies from one place to another, e.g. up to 100 mm in coastal areas, 350 mm on highlands, and only 30 mm in the desert further south.

The second climatic region includes the Dhofar Mountains and the adjacent coastal plains to the south. Temperatures are relatively moderate throughout the year and rarely exceed 35°C in summer and 17°C in winter. Rain falls as gentle drizzle during the *khareef* (season of monsoon rains) from June to mid September, and as heavy showers in April–May and October–November. The rainy part of the Jabal is largely confined to the area south of the watershed, and the monsoon clouds extend very little beyond this point. Aridity increases rapidly north of the watershed; the northern slopes lie in the rain shadow area. Thamrait, a town north of the mountains, shows typical desert climatic conditions. Trees, shrubs and tall grasses play an important role in maximising the amount of effective precipitation received from the monsoon mists as they roll over the plateau (IRI 1976). About 100 mm are received annually in the Salalah Plain, while in the mountains a few kilometers to the north, up to 500 mm may be received. There are great variations from one place to another (Sale 1980a). The highest average of 300 mm is received at Jabal Al-Akhdar and in some parts of the Dhofar Mountains, followed by Oman Interior (South of Oman Mountains) receiving about 200 mm annually. The Batina and Salalah Plains receive about 100 mm, whereas Shargia receives only about 90 mm (Hammad 1981). Some parts of the deserts hardly receive any rains at all.

Thus, the plains below 500 m receive very little rain to sustain agriculture which depends entirely upon underground water. The numerous wadis drain the mountains and maintain the level of underground water. A programme is currently being carried out to build dams in all the important wadis to increase the amount of water stored and prevent water from being wasted into the sea.

AGRICULTURE

It is estimated that there are about 100,000 ha of good agricultural land in Oman (Hammad 1981), most of which is utilized in crop production. Other potential areas are also available. In addition, there are vast rangelands on which large numbers of livestock are raised. The most important crops grown in Oman are date-palm, mango, lime, banana, coconut, alfalfa, vegetables and wheat. Grains such as sorghum, maize, barley and pearl millet are also grown, mainly as fooder crops. For more details about agriculture in Oman, see Hammad (1981).

LIVESTOCK

GRM International (1984) estimated the number of livestock in Oman as shown in Table 1.

Cattle and goats are considered the most important species, while camels are not fully utilized despite their large numbers. The Jabali cattle of the Southern Region, having Zebu characteristics, are larger and more vigorous than those of the North which have the characteristics of *Bos taurus*. During the *khareef* (monsoon), cattle in Dhofar Mountains are kept indoors during the day to avoid the biting flies, and are allowed to graze only at night. Food supplements, such as processed pellets and dried sardines, are used during the dry season, particularly in the Southern Region. Goats are the predominant species in the north, being mainly concentrated in the Batina Plain and the Muscat Region. In the Southern Region goats resemble the larger Somali type with black head and white or brown coats. Sheep are few

Table 1. Number of livestock on Oman.

Region	*Cattle*	*Goats*	*Sheep*	*Camels*
Batina and the capital area	15,304	179,915	60,799	2,872
Masandam	272	26,537	3.083	162
Western Hajar	4,905	44,905	29	1,469
Eastern Hajar	2,426	24,361	49	792
Buraimi	1,991	31,652	4,004	2,956
Dhahira	7,975	90,447	14,918	3,380
Oman Interior	7,505	88,056	12,491	1,165
Shargia and Ja'alan	7,856	104,074	18,081	510
Southern Region	77,922	106,249	3,977	54,365
Total	125,886	696,199	135,654	71,488

in number in comparison to goats, and those found in the Southern Region are of Somali origin. Camels are found mostly in the Southern Region. They are relatively small and are used for milk production and racing.

Serious infectious diseases are under control in Oman, thanks to the efficient veterinary service available. National vaccination campaigns cover rinderpest, foot and mouth, blackleg and pox diseases, etc.

RANGELANDS

The 700–800 km-wide strip of desert between the Dhofar Mountains and the Oman Mountains is reckoned to be an effective plant barrier (Lawton 1978) as some plant species are only found in one range of mountains and not in the other. Lawton (1978) maintains that it is likely that Jabal Al-Gara in Dhofar is the eastern limit of many species of African affinity, and Jabal Al-Akhdar in Oman Mountains is the western limit of some Asian species. Therefore it seems logical, due to differences in plant species and climatic conditions, to deal with rangeland management in the two regions separately.

OMAN MOUNTAINS RANGE AND THE SURROUNDING AREAS (NORTHERN REGION)

Studies of natural vegetation in this region are few in comparison with those in Dhofar Mountains. One of the most important recent studies is that of Mandaville (1977). It was carried out in order to determine which of the Jabal Al-Akhdar plant communities may play an important role as a food source for herbivorous mammals, particularly the 'tahr' (*Hemitragus jayakari*). Mandaville described the vegetation in central Jabal Al-Akhdar and divided it into five broad zones: (i) *Acacia* desert parkland, (ii) Mountain wadi associations, (iii) *Euphorbia larica* shrub slopes, (iv) *Reptonia-Olea* woodland, and (v) Juniper summit zone. He also described the plant communities in the Jabal Aswad Region (east of Jabal Al-Akhdar) which were found to parallel those of Jabal Al-Akhdar but with significant differences in some aspects. Mandaville (1977) noted that in central Jabal Al-Akhdar, plant communities were not threatened by excessive grazing pressure from domestic livestock or wild species.

Studies on vegetation in other parts of this region such as the Batina Plain, Shargia, Oman Interior and Dhahira were not undertaken, and the need for such studies cannot be over-emphasized. However, rangelands in these areas show signs of deterioration, though to a less serious extent than in the Dhofar Mountains rangelands (GRM International 1984). The factors responsible for this deterioration are: (i) Depleted soil seed reserves, (ii) Low moisture level leading to poor regeneration, and (iii) Excessive grazing pressure which prevents successful regeneration. Long-term availability of cheap forage for a viable livestock industry requires urgent work to reverse these trends. GRM International (1984) suggested that rangelands in this region can be improved by planting areas with drought-tolerant indigenous and exotic species and that these areas should be set aside as seed and forage reserves. By siting these reserve areas along the wadis, they would provide seed for deposition downstream for increased natural regeneration of the wadi-fan areas.

A long-term research programme is required to investigate and solve problems to ensure the successful improvement of these natural rangelands. Successful plant species may include trees, shrubs, grasses and legumes. May (1981) discussed the role of tropical legumes in marginal lands improvement, and many of the species he mentioned may be of possible use in Oman's marginal lands.

Another possible way for improvement of the rangelands in these areas is the regeneration of aquifer recharge zones (GRM International 1984) by utilizing pond flood water behind the embankment and controlling water release through culverts. The reduced volume and velocity of water passing through the culverts will increase water infiltration in the zone immediately downstream of the embankment. It is this zone which may be exploited in rangeland regeneration. This type of regeneration may be applicable to the Wadi Al-Khawd dam which has a 4.9 km long and 8 m high embankment, breached at its base by eleven 1.2 m pipe culverts. An estimated area of 750 ha can be improved around the dam. This technique, however, is most suited to the upper reaches of wadis in the Oman Mountains and Masandam Regions rather than coastal plains. Other wadis which may be considered for improvement are Dunk, Nizwa, Samail, Tayin and Bani Khalid. However, the current aquifer recharge programme is directed towards the coastal plains where the need for water conservation is most urgent to meet the increasing domestic and agricultural requirements.

In the Batina Plain, due to steady expansion of irrigated agriculture, steps were taken to reserve areas as natural rangelands. These areas are the North Batina and the South Batina. In the North Batina (4250 ha), a 5-ha nursery, costing about 200,000 Omani Rials (US$520,000) will be established for production of grass and legume seed as well as seedlings of shrubs and trees for regeneration of the reserved rangelands. In the South Batina, an area of 170 km^2 is reserved and it is estimated that the cost of regeneration will be about 110,000 Omani Rials. In addition, 350,000 O.R. were allocated to establish a 50 ha area in the desert, with the objective of investigating and studying the development and improvement of natural rangelands (Ministry Agriculture and Fisheries 1984).

THE DHOFAR MOUNTAINS (SOUTHERN REGION)

Many studies were undertaken in this region concerning vegetation, climatic conditions as well as other aspects relevant to rangelands (IRI 1976, Lawton 1978, Sale 1980a and b, Radcliffe-Smith 1980, Mandaville 1980, Zaroug 1983, Anonymous 1983, GRM International 1984). It is widely recognized that the mountain range of Dhofar is an ecologically unique area of the Arabian Peninsula (Anonymous 1983). As the winds of the southwest monsoon reach the Dhofar coast, the warm moist air is cooled, forming mist and cloud. The Dhofar hills are limestone which is more water-retentive than the siliceous rocks, and this would encourage, under humid conditions, the development and maintenance of a good vegetation cover (Radcliffe-Smith 1980). Lawton (1978), Sale (1980b) and Radcliffe-Smith (1980) described the ecology and vegetation of the Dhofar Mountains, while Zaroug (1983) reported on rangeland conservation and development.

Sale (1980b) divided the region into five ecological zones based mainly on rainfall and soil type. These zones, together with their estimated areas, are: (i) Desert (97,000 ha), (ii) Semi-desert scrub (77,500 ha), (iii) Short grassland (90,000 ha), (iv) Tall grassland (55,000 ha), and (v); Woodland (52,000 ha). Clearly, the most important of these zones are the grassland and woodland zones, and together they represent the best rangelands in Oman and probably in the whole of Arabia. The sequence of zones from north to south is very clear.

Radcliffe-Smith (1980) gave detailed description of the plant communities including lists of plant species. The vegetation is composed of two elements: (1) In the green parts the Sudano-Deccanian element is strongly represented with affinities westward to Yemen, Ethiopia and African savannas and eastward to peninsular India; (2) In the most arid areas to the north of the Jabal, it is the Saharo-Sindian element which predominates, with affinities to the vegetation of the deserts of North Africa and northwest India. Radcliffe-Smith (1980) also recognized eight environmental zones in Dhofar, namely:

(1) The littoral belt.
(2) The cultivated belt, mainly in the vicinity of Salalah.
(3) The Salalah Plain.
(4) The southward-draining wadis of the Jabal.
(5) The south-facing slopes and the cliffs of the Jabal.
(6a) The grassland on top of the Jabal.
(6b) The Jabal to the west of Hasik.
(7) The upper dip-slope immediately behind the crest of the Jabal.
(8) The lower dip-slope, dipping gradually towards the empty Quarter.

The Jabali people, estimate at about 20,000, have grazed the plateau for more than 2000 years (Lawton 1978) and evolved a pattern of land use that has proved to be in balance with their environment. The Jabal grasslands, together with the Salalah short grassland to the south, support herds of cattle, while the semi-desert scrub-grassland to the north supplies herds of camels and flocks of goats with their feed requirements (Zaroug 1983). Crops of sorghum, cowpea and millet were also grown during the *khareef* but the practice had been abandoned in the last few years. Stone-walled cultivation plots are still present.

The Jabali people are semi-nomads, and their movements together with their herds are governed by season and forage availability. This movement between *khareef*, winter and summer is in such a manner that some parts of the pastures are retained as standing hay for dry season use. This concerted movement of man and animal was disturbed by a number of factors due to the impact of developmental changes in the region, especially the construction of a surfaced road system, improved medical and veterinary services, improved water supplies through boreholes in the Jabal and the Sahl, and use of vehicles for off-track transport. The authorities are well aware of the potential environmental hazards of these inevitable developmental changes; and a ministerial committee has been set up to co-ordinate development in the Southern Region, and to monitor its impact on the environment to ensure harmony between development and the environment (Royal Decree No. 48/84).

The deterioration of rangelands in this region can be attributed to several factors: (i) The limited amount of rainfall and its irregular distribution; (ii) A sharp increase in the number of livestock; (iii) The construction of a modern road system and its associated unsurfaced off-tracks; (iv) The dominance, in grasslands, of short-lived annual species and the gradual disappearance of perennial plant species; and (v) A change of plant species from palatable to unpalatable or even poisonous species, particularly around water points. Other possible reasons could also be cited, and in combination all factors are responsible for the present state of the rangelands. However, these trends can be reversed and improvement can be achieved.

(i) Rainfall

There are no long-term records to show the trends of rainfall in the region. The annual precipitation, however, falls short of the requirements for development (Anonymous 1983). Furthermore, there is great variability in the amount and distribution of precipitation within the *khareef* and other seasons of the year which may affect grassland productivity and its floristic composition (Zaroug 1983). Off-season rain storms cause floods and soil erosion in areas depleted of vegetation, e.g. around watering points. Drought conditions are also experienced in some years (Zaroug 1983) due to either poor rains or delay in their start. Although it is extremely difficult to do anything about rainfall distribution, a number of techniques can be used to increase the amount of precipitation. Cloud seeding technique has been advocated (Anonymous 1983) especially during *khareef* (late June–mid September) and winter (December–March) where conditions may be favourable. Another approach to increase the amount of precipitation is through the use of vegetation. Evidence obtained by IRI (1976) suggests that shelterbelts, oriented to catch monsoon winds, increased precipitation by about five-fold. The vegetation condenses monsoon mists and recharges the underground water that feeds the springs and boreholes (Lawton 1978). The structure of the vegetation is also important in this condensation process; trees, shrubs and tall grasses will condense more mist than shorter types of vegetation. According to Lawton (1978) the dwarf shrub *Euphorbia balsimifera* subsp. *adensis* is essential for the maintenance of natural pasture in the semi-desert (Najd ara) north of the Jabal. Any rangeland improvement must take into account the role of trees, shrubs and tall grasses. Fodder trees and shrubs can be planted as hedges and windbreaks on the Jabal.

(ii) Livestock numbers

The sharp increase in the number of animals is due to a few factors: (a) the improved veterinary service rendered available by the Animal Wealth Department; (b) the increased income, from various sources, that allows herdsmen to buy processed feed and baled hay, and become very reluctant to sell surplus unproductive animals; and (c) a possible increase of the annual calfing rate due to better nutrition and veterinary services.

(iii) Construction of a modern road system

Construction of a modern road system is undoubtedly a commendable development which opens remote areas for government services. Lawton (1978) dealt with the impact of the construction of all-weather roads and the provision of boreholes on the grazing land ecosystem. He suggested that vegetation should be re-established in areas disturbed by road construction as soon as possible, in order to prevent possible soil erosion by wind and water. The problem of multiple off-tracks remains to be solved, and the recent phenomenon of dust storms in the Jabal in winter is attributed to vegetation damage and exposure of soil by vehicles using these tracks.

(iv) Change of floristic composition

Zaroug (1983) presented evidence suggesting that a change in the dominance of plant species occurred between the 1981 and 1982 monsoon seasons. In the Tawi Atair area *Impatiens balsamina*, an unpalatable species, covered about 48% of the area in 1981. During the monsoon of 1982, a tall unpalatable legume known locally as 'mahlab' constituted about 50% of the vegetation cover in the same site, while *Impatiens balsamina* formed only 10%. He identified a toxic legume, *Crotalaria retusa*, in the same area, and noted a degradation of range vegetation for about 2 km around the watering point. Zaroug's field observations also indicate that there was a shift from tall grasses to short grasses or unpalatable broad-leaf species. Uninterrupted grazing by a large number of animals prevents both annual and perennial grasses from forming sugars and carbohydrate needed by the annuals for growth and seed formation, and by the perennials for regrowth after their dormancy period during the dry season (Zaroug 1983).

A number of steps were taken to arrest the serious deterioration of rangelands in the Southern Region of Oman. They include:

(1) A destocking programme; decreasing the number of animals on these rangelands is the first step towards their improvement. This must be accompanied by a livestock marketing system with regular and reliable outlets to syphon off the excess numbers of animals. Research elsewhere (Poollen and Lacey 1979) indicates that moderate stocking is the best way to realize optimum economic returns as well as conserve the rangeland. The Ministry of Agriculture and Fisheries' subsidy programme, which began in 1984, offers animal owners 1.5 O.R.

(US$3.9) per kg live-weight. GRM International (1984) recommended that cattle numbers be reduced from 70,000 in 1984 to 50,000 in 1990.

(2) The second important project to improve rangelands in the region over the next five years, at an estimated cost of 5.4 million O.R., consists of three phases as follows (Ministry of Agriculture and Fisheries 1984):

(a) The reseeding of 180,000 ha in Dhofar Mountains, of which 70,000 ha are open lightly wooded grasslands and the remainder comprise woodlands and escarpments. Aerial seeding as well as other methods will be used, two to three weeks after the start of the *khareef*. The following pasture and browse plant species are recommended for this project:

(i) legumes: *Stylosanthes humilis;*
(ii) grasses: *Apluda mutica*, *Cenchrus ciliaris*, *C. setigerus*, *Themeda quadrivalvis*, and *Urochloa mosambicensis;*
(iii) browse species: various *Acacia* spp.

For specific areas in the woodlands, *Clitoria ternatea* and cowpea (*Vigna unguiculata*) are also recommended.

(b) The selection of 19 5-ha sites in different parts of the Jabal, near the main tribal settlements. These sites should be used as fenced demonstration and emergency pastures. Rows of trees, shrubs and tall grasses will be grown to condense the monsoon mists. The following plant species will be grown in these reserved areas: *Acacia tortilis*, *A. siliqua*, *A. salicina*, *A. victoriae*, *A. cyclops*, *A. armpegiana*, *A. senegal*, *Chloris gayana*, *Ciricidia maculata*, *Pittosporum phylliraeoides*, *Anogeissus dhofarica*, *Conocarpus lancifolius* and *Zizyphus spina-christi*. It is hoped that by growing the trees and shrubs they will condense the mist, and the grasses and legumes growing underneath will take advantage of the increased amount of precipitation thus made available. Protection of these areas is important as the species planted by the International Research Institute in 1975 had disappeared by 1981 when the fence was removed (Zaroug 1983).

(c) A 100-km^2 area in the Salalah Plain between Razat and Taqa will be planted with drought-tolerant browse trees and shrubs mainly for camel grazing. Nurseries will be established to supply grass and legume seed as well as seedlings of trees and shrubs.

IRRIGATED FORAGE PRODUCTION

Research work on irrigated forage crops is carried out in research stations at Salalah (Southern Region), Rumais (Batina Plain) and the Oman Interior. These forage crops include alfalfa, Rhodes grass, Napier grass, pearl millet, sorghum and maize as well as various legumes such as cowpea, pigeon pea, vetches, pea, *Dolichos lablab* and clitoria. No attempt is made in this report to review and summarize the results obtained so far, but further details can be obtained from the annual reports of these stations. This research effort is mainly directed towards increasing forage production under irrigation where mixed farming is of primary importance. A good example of this mixed farming is illustrated by the Al-Khabura Development Project Farm on the Batina coast. The work is undertaken by the University of Durham, England in co-operation with the Ministry of Agriculture and Fisheries, Oman, and supported, since its inception, by Petroleum Development, Oman. The work has involved trials on irrigation, forage crops, small livestock husbandry and bee-keeping. The goat is promoted as a dual purpose animal for both meat and milk production.

CONCLUSIONS

Observations and results from many studies concerning rangelands in Oman show that deterioration had occurred in different parts of the country, but it varies in degree and extent from one area to another. Measures are taken to correct the situation, and these have been discussed in appropriate sections of this report. However, it is clear that in order to improve rangelands, as a source of cheap fodder for the livestock industry, research must be given top priority. The current trend in the Ministry of Agriculture and Fisheries to create units of research, under the umbrella of the existing Agricultural Research Organization, to tackle problems involving forestry and rangelands as one inter-related discipline is highly commendable. There is already one such unit in Salalah. Long-term improvements in rangelands, forestry, de-desertification and other related aspects require strong local research programmes, executed under the local environmental conditions. Foreign experience, valuable as a guide line, will never be a substitute for locally acquired experience. Therefore it is suggested that rangelands and forestry improvement should be tackled on a national level with strong regional programmes. Such a set-up, in which a national co-ordinator will play an important role, will ensure that urgent national and regional problems will be resolved. A parallel extension service is a prerequisite for the successful application of research results in the field.

Co-operation with other Arabian Gulf States is essential in order to exchange ideas, experience and material.

ACKNOWLEDGEMENTS

We are grateful to Professor Ahmed Babiker and Dr Mohamed El-Noor for helpful comments on the manuscript, and also to the latter for providing the information on which the section on livestock is based. Thanks are also due to many colleagues in the Ministry of Agriculture and Fisheries for their help, especially Mr Pasha Badawi.

LITERATURE CITED

Anonymous. 1983. Recommendations of the Salalah Workshop on environmental aspects of development in the Southern Region. Diwan of Royal Court Affairs, Salalah, Sultanate of Oman.

GRM INTERNATIONAL. 1984. A master plan for development of the livestock sector. Unpublished report. Ministry of Agriculture and Fisheries, Muscat, Sultanate of Oman.

Hammad, A. E. 1981. The development of agriculture and fisheries in the Sultanate of Oman (Arabic text). Ministry of Agriculture and Fisheries, Muscat, Sultanate of Oman.

International Research Institute (IRI). 1976. Summary of results. Unpublished report. Ministry of Agriculture and Fisheries, Muscat, Sultanate of Oman.

Lawton, R. M. 1978. A reconnaissance survey of the Jabal Qara grazing ecosystem, with particular reference to the impact of development. Unpublished report to the Sultanate of Oman. Ministry of Overseas Development, London.

Mandaville, J. P. Jr. 1977. Plants. *Journal of Oman Studies,* Special Report No. 1: 229–267.

Mandaville, J. P. Jr. 1980. Frankincense in Dhofar. Journal of Oman Studies, Special Report No. 2: 87–89.

May, R. M. 1981. Useful tropical legumes. *Nature* **294:** 516–517.

Ministry of Agriculture and Fisheries. 1984. Programme of work to execute the Royal directives concerning agricultural lands, natural rangelands and fisheries. Unpublished report (Arabic text). Ministry of Agriculture and Fisheries, Muscat, Sultanate of Oman.

Poollen, H. van Walt and J. R. Lacey. 1979. Herbage response to grazing systems and stocking intensities. *J. Range Manage.* **32:** 250–253.

Radcliffe-Smith, A. 1980. The vegetation of Dhofar. *Journal of Oman Studies,* Special Report No. 2: 59–86.

Sale, J. B. 1980a. The environment of the mountain region of Dhofar. *Journal of Oman Studies,* Special Report No. 2: 17–23.

Sale, J. B. 1980b. The ecology of the mountain region of Dhofar. *Journal of Oman Studies,* Special Report No. 2: 25–54.

Zaroug, M. G. 1983. The status of the rangelands of the Southern Region of the Sultanate of Oman (Dhofar) and prospects for their conservation and sustained development. A report presented to the Salalah Workshop, 22–30 September 1983. FAO, Rome.

Range improvement of the semi-arid regions by introduction of some forage crops

LAIKA K. AL-JIBURY

Council for Scientific Research, Agriculture and Water Resources Research Center, P.O. Box 255, Al-Jadriyah, Baghdad, Iraq

ABSTRACT

The continuous decrease of natural forage production under increased grazing was observed and quantified in most parts of the desert rangelands of Iraq. Most of the palatable and nutritious plants of the Iraqi ranges have disappeared, resulting in lower livestock production. There is an urgent need, therefore, to develop new agricultural systems in the marginal cropping areas of the Jezira, where farmers adopt a fallow system, cultivating cereals for one year and allowing the land to rest for one or two years.

The introduction of some annual legumes in the area becomes important in order to provide adequate fodder for the livestock in the region. A research project was designed to find out information on relative productivity of some forage legumes in the marginal cropping areas of Iraq.

The results show that some species can provide fodder of good quality. Consequently, recommendations are made to the farmers to grow *Vicia sativa*, *V. narbonensis* and *Medicago scutellata* in the Jezira under rainfed conditions, using a crop-legume rotation system instead of the traditional crop-fallow system.

INTRODUCTION

Iraq has a surface area of about 440,000 km^2, of which about 50% is considered as rangelands. These are located in the Jezira area and the Western and Southern Deserts. It is estimated that 90% of the forage comes from these desert rangelands.

The absence of a national grazing policy has led to the misuse of the desert rangelands through various practices. Consequently, productivity of these rangelands has decreased considerably because of overgrazing and fuel gathering. As a result, large areas of the desert shrublands of the country are severely damaged and sometimes abandoned (Thalen 1979).

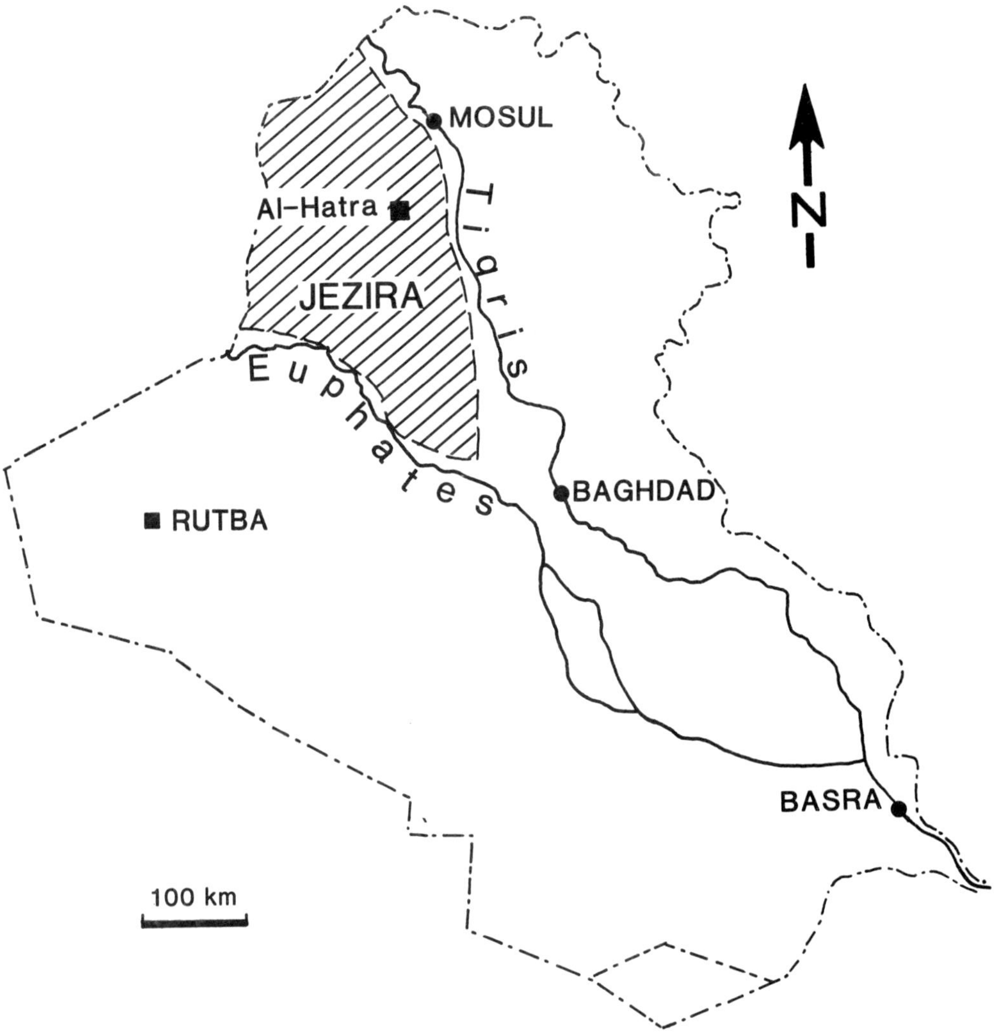

Fig. 1. Map of Iraq showing the study area of Al-Hatra.

Consequently, natural regeneration of a productive vegetation has become difficult or even impossible over extensive areas.

The Jezira region, an area in the northwest of Iraq between the Tigris and Euphrates rivers (Fig. 1), was an excellent grazing land in the 19th and the first half of the 20th centuries (Ainsworth 1941). The spread of dense settlement in this part of the country resulted in intensive grazing, ploughing and cultivation in combination with destructive removal of perennial shrubs for fuel. This has led to the gradual disappearance of valuable species.

Efforts are now being made towards improvement, conservation and development of the degraded rangelands. In 1963, the Iraqi authorities delineated the southern boundary for rainfed agriculture. Accordingly, crop cultivation (wheat and barley) at Al-Hatra region (Fig. 1) has been prohibited in the context of a management practice and the region was assigned solely for improvement and development of range resources.

A research project was initiated in 1978 to survey the natural vegetation cover of Al-Hatra area. Investigations revealed that due to the abuse of the native vegetation, natural plant succession has been disturbed. As a result, poisonous and non-palatable plants dominated the area, with few edible species for livestock.

Considering the importance of forage crop production in the Jezira and the need for livestock improvement and management, a new forage production system was proposed for the marginal cropping areas of the Jezira (Al-Jibury and Al-Hadeethy 1984), in order to augment forage production along with cereal production (wheat and barley) under rainfed conditions. The term marginal cropping areas is used to designate the boundary between the cultivated areas and rangeland. The areas cover all lands where the average annual rainfall is between 200 and 250 mm. Below this amount, rain is inadequate for economic crop production.

Rainfed areas in Iraq are monocropped, where farmers cultivate in November and harvest in May, then keep the land fallow for one or two years. The introduction of winter legumes could play an important role in the region in providing sufficient forage for livestock (Haydar 1976). Moreover, the winter legume cover also protects the soil against erosion, and improves the physical conditions of the soil (Purchase *et al.* 1949). The nodule bacteria on the legume roots increase the nitrogen content of the soil.

THE PRESENT STUDY

Five annual winter legumes were chosen for sowing in the region at the beginning of the rainy season. They include *Vicia narbonensis* (narbon vetch), *V. sativa* (common vetch), *Medicago scutellata* (snail medic), *M. truncatula* (barrel medic), and *Trifolium subterraneum* (subclover). The investigation was continued for three successive years (1980/81–1982/83) with the objective of estimating the productivity of these legumes in the marginal cropping areas of the Jezira.

RESULTS AND DISCUSSION

The results presented in Table 1 show that, in general, forage production of the cultivated legumes during the three successive years is economically feasible. Table 1 shows also that productivity of each species differs from one year to another. These variations in forage production may be related to the fluctuations in the amount of rainfall and/or the differences in time and distribution of rain throughout the growing season (Table 2). Table 1 also reveals that common vetch and narbon vetch were more productive than medics. Productivity of snail medic was higher than that of barrel medic. Subclover showed very poor growth during the experimental period, and hence data on its performance were not included in Table 1.

Table 1. Forage production under rainfed condition of the marginal cropping areas of the Jezira (1980–1983).

Legume species	*Amount of green forage (ton/ha)* 1980/81	1981/82	1982/83	Average
Common vetch	5.58	5.80	7.43	6.27
Narbon vetch	8.25	6.31	9.00	7.85
Snail medic	4.62	3.32	9.60	5.84
Barrel medic	3.57	1.37	3.87	2.94

Table 2. Monthly rainfall in the marginal cropping areas of the Jezira 1980–1983.

Month	*Rainfall (mm)* 1980/81	1981/82	1982/83
October	25.4	1.2	12.0
November	46.2	29.9	60.2
December	36.7	40.4	58.3
January	8.8	58.5	47.2
February	36.1	45.2	36.8
March	26.1	1.2	29.3
April	50.9	18.2	48.8
May	0.5	10.3	3.3
Total	230.7	204.9	295.9

Accordingly, recommendations were made to the farmers to cultivate the three forage crops: narbon vetch, common vetch and snail medic, in the marginal cropping areas along with cereal crops, using a crop-legume rotation system instead of the traditional cereal-fallow system.

LITERATURE CITED

Ainsworth, W. 1941. Notes of an excursion of Kal'ah Sherkat, the U'r of the Persians, and to the ruins of Al-Hadr, the Hutra of the Chaldees, and Hatra of the Romans. *J. Roy. Geog. Soc.* **11:** 1–20.

Al-Jibury, L. and A. Al-Hadeethy. 1984. Range improvement of the semi-dry region (Jezira), northern part of Iraq, by introduction of some fodder crops in the rainfed area of Talaafar. *J. Agric. Wat. Resources Res. Center, Baghdad* **3** (2): 97–103.

Haydar, A. H. 1976. The importance of legumes in crop rotation at the rainfed area, northern part of Iraq. The Third Arab Symposium for Field Crops, Baghdad, 8–13 May 1976.

Purchase, H. F., J. M. Vincent and L. M. Ward. 1949. The contribution of legumes to soil nitrogen economy in New South Wales. *J. Austral. Inst. Agr. Sci.* **15:** 112–117.

Thalen, D. C. P. 1979. *Ecology and utilization of desert shrub lands in Iraq. Dr W. Junk, Publishers, The Hague, 428 pp.*

Forage trials under Kuwait's conditions

FAISAL K. TAHA, SAMIRA A. OMAR AND AHMAD A. NASSEF

Aridland Agriculture Department, Food Resources Division, Kuwait Institute for Scientific Research, P.O. Box 24885, Safat, Kuwait

ABSTRACT

Forage supplied by rangelands is limited due to seasonal and erratic rainfall, high temperature, high evaporation rates and fresh water shortage. This is clearly demonstrated in the importation of green forage (70%) and concentrate feed for livestock (100%).

Research trials for forage production were concentrated on seed germination and seedlings establishment of native and exotic species, use of saline water for *Atriplex* seedlings establishment and growth of desert halophytes under rainfed systems. Of the tested species, the promising ones were *Cenchrus ciliaris*, *Pennisetum setaceum*, *Eragrostis curvula*, *Sporobolus airoides*, *Agropyron riparium*, *Atriplex canescens*, *A. halimus* and *A. lentiformis*. The results also suggest the possible use of available saline water of 5,000 mg l^{-1} (total dissolved solids) for seedlings establishment of *Atriplex* species. Saline water was applied following plant germination and was continued for 12–15 weeks without adverse effects.

Atriplex species grown under Kuwait's natural rainfall (110 mm) showed initial positive response. However, with the onset of summer and the occurrence of high temperature extremes and high radiation, and the insufficient moisture, the plants were adversely affected, with *A. halimus*, *A. canescens* and *A. glauca* reduced to 9, 15 and 56% survival, respectively.

Current emphasis is to further investigate seed viability of native species for large-scale planting as well as expanding research of desert halophyte transplanting and use of supplement irrigation during initial establishment.

INTRODUCTION

In the production of forages in Kuwait, major limitations are the shortage of rainfall, harsh weather with prolonged hot summers, sandy soil with very low fertility and lack of permanent surface water flow. These constraints have resulted in limited forage supply and the country's dependence on the importation of green fodder. Although the environmental constraints impose great difficulties for forage production, they are not insurmountable. With today's technology and know-how in desert agriculture and the proven results from various range improvement practices, it is possible to significantly increase the production of fodder crops. A feasible approach for Kuwait to increase its forage production seems to be the exploitation of native plants as well as supplementing the forage supply with additional resources through the intoduction of new species suitable for the environment. Realizing the severe shorage of fodder for the livestock industry in Kuwait, a major component (task) of the on-going management activities (Taha *et al.* 1984a and b) has been devoted to this purpose.

Research to circumvent problems encountered in growing native desert vegetation has centred on examining conditions favouring germination and seedling growth with major emphasis on the optimal sowing time for the plants' emergence and establishment. This is particularly important at present when re-establishment of the country's desert vegetation is a key factor to the proper management and planning of its rangelands (Plummer and Fenley 1950, McGinnies 1960, Kernick 1964, Taha and Omar 1980, Mott 1977). Increase of forage production by introducing exotic species is considered to be of the utmost importance (Dowling *et al.* 1971, Plummer and Fenley 1950, Springfield 1970, Omar and Taha 1981, Behbehani *et al.* 1981). The shortage of native species seed supplies, caused by overgrazing, and the perennial shortage of green fodder, which is critical in the summer months, necessitates the extension of the grazing season by introducing new species. Thus, germination tests of exotic species were conducted under simulated Kuwait environments.

Depletion of forage is evident in many areas of Kuwait's rangelands, and to bring these areas to optimum production will require some type of artificial regeneration. This will necessitate the establishment of a good vegetation stand that is salt tolerant, heat resistant and protected from grazing pressures. One feasible avenue is through transplantation, since direct seeding of many range species in areas receiving only 100–150 mm of rainfall annually is difficult and unreliable (Nemati 1977, Goebel 1978). *Atriplex* species appear suitable for transplanting in Kuwait because of their resistance to drought, salinity and defoliation and hence greenhouse trials and field investigations were conducted in an attempt to successfully establish them under Kuwait's conditions.

The present paper will attempt to summarize current research efforts on forage production. For simplicity and clarification, the discussion will be divided into three sections: (1) seed germination and seedling establishment of native species; (2) research trials on exotic species; and (3) establishment of desert halophytes using brackish (saline) water under natural field condition.

METHODOLOGY

1. *Description of experimental design for native species:* Seeds of native species were collected from various ecologically important zones of

Kuwait's desert plant communities. They were refined manually and by sieving. Testing for seed germination of native species depended on the quantity of seeds available for individual species. In all treatments, tests were conducted on seeds which were still retained in the utricles. This was intended to allow the study of the seed's germination behaviour in a manner similar to that which occurs in nature.

Experiment conducted under simulated environmental conditions: Treatments involved the simulation of climatic conditions (temperature and relative humidity for selected calendar months of the year. Data used were averages of 25 years climatological information in Kuwait. Simulation for individual months were made possible by the use of a Seed Germinator (Model G30, a product of Controlled Environments Inc., Winnipeg, Canada) which was programmed to simulate temperature and relative humidity parameters according to desired monthly information. A light intensity of 160 foot candle was set during the day, and day and night lengths were also monitored. Monthly data used are summarized in Table 1. The seeds were incubated for 20 days and were

Table 1. Description of the simulated environmental conditions used for seed germination of some native plants.

	*Simulated environmental conditions**			
*Month**	*Max. Temp. (°C)*	*Min. Temp. (°C)*	*Max. R. Hum. (%)*	*Min. R. Hum (%)*
February	20.7	9.3	80	31
March	26.1	13.5	69	23
April	31.2	18.3	66	22
May	38.2	23.7	65	19
October	35.5	19.5	65	19
November	26.5	13.9	77	31
December	20.0	8.5	70	70

*Data are averages of 25 year's (1957–1972) climatological information from Kuwait International Airport.

considered to have germinated when the plumule was longer than the seed and was visible to the naked eye. Germination counts were made daily. For each individual plant species, 50 flower pots were filled with salt-free vermiculite (used as a soil mix) and planted with five seeds. Pots were placed on stainless steel trays, filled with 500 ml of distilled water and were placed in the seed germinator.

2. *Description of experimental design for exotic species.*
   (a) Seed germination under six environmental conditions: Species were tested under six environmental conditions. These environments are described in Table 2 and were programmed in the seed germinator. Environmental parameters included maximum and minimum temperatures, maximum and minimum relative humidity for day and night as well as day length. A light intensity of 160 foot candle was set during the day for all environments. Exotic species were grouped

Table 2. Description of the simulated environmental conditions used for germination of exotic species.

Environment no.	*Max. Temp. (°C)*	*Min. Temp. (°C)*	*Max. R. Hum. (%)*	*Min. R. Hum. (%)*
1	45	33	91	29
2	30	20	70	30
3	33	10	88	35
4	26	14	74	34
5	30	14	63	40
6	38	26	80	55

according to type (shrub, forb or grass) and the groups were tested under all the environments. Germination tests for individual species were conducted without prior seed treatment.

(b) Another experiment conducted on exotic species was to simulate the Kuwait climatological data shown in Table 1 for the months of February, March, April and October. Similar procedures were utilized for data collection of germination tests.

(c) Seed germination in different soil media: Another group of the exotic plant species were tested in the greenhouse during the month of October. Plants were tested in different soil media (peatmoss and agricultural soil mix of 1:1 ratio, and 100% peatmoss). Seeds were placed in paper flower pots and then placed in wooden boxes inside the greenhouse. Thirty-five to 70 paper flower pots were used for each plant species, and treatments were either replicated or triplicated based on seed quantity.

3. *Establishment of desert halophytes utilizing brackish water and dry farming practices*

(a) *Use of saline water on desert shrubs.*

The present experiment was conducted to investigate the effect of saline water available in Kuwait on the genus *Atriplex*. The study included various irrigation treatments as well as many *Atriplex* species.

(i) *Experiment I – Atriplex halimus:* In this experiment, *Atriplex halimus* was used and the irrigation treatments were nine (sewage water, brackish water, seawater dilutions of 1:15, 1:7, 1:3, 1:1, 3:1, 1:0 and control). Twenty seedlings, four weeks old, were assigned to each treatment. Pots were arranged in a random design, and were placed in a KISR greenhouse having average maximum and minimum temperatures of 34.7 and 26.2°C. Relative humidity recorded was 63.8%. Irrigation treatments were applied every 14 days. Measurements of plant height and width were taken once every two weeks and climatological parameters of temperature and relative humidity were recorded.

(ii) *Experiment II – Atriplex halimus, A. glauca and A. canescens.* Data obtained from the previous experiment on the effect of saline water on the growth of *Atriplex halimus* showed that sea dilutions between 1:15 and 1:7 were suitable for seedling establishment. This experiment was carried out as a follow-up to investigate the effect of various dilution levels between 1:15 and 1:7, such as 1:13, 1:11, 1:9 and control, to find the optimum dilution level for seedling growth. The experiment also tested three shrub species: *Atriplex halimus, A. glauca* and *A. canescens*. Twenty seedlings, two weeks old, of each species were assigned to each treatment and were arranged in a random design. Irrigation treatments were applied every 14 days. Measurements of plant height and width were taken at bi-weekly intervals and climatological parameters of temperature and relative humidity were recorded.

(b) *Dry farming experiments*

Shrub transplanting under dry farming practices has been regularly used in revegetation trials of depleted rangelands (Knowles and Condon 1951, Plummer *et al.* 1966, Goodin and McKell 1970, Chatterton *et al.* 1971). This method has been widely used in arid areas (Springfield 1970, Nassef *et al.* 1981, Taha *et al.* 1983).

A preliminary experiment was carried out to investigate the performance and growth of three *Atriplex* species and one *Simmondsia* species under Kuwait's natural conditions without the application of irrigation water. Specific objectives were to obtain information on plant production parameters (height and width) as well as percentage survival under these harsh conditions. An area of 440 × 230 m was selected at the Institute's Sulaibiya site for dry farming practices. It was surveyed and contour lines were marked. The area was ploughed and embankments of 25–30 m width were constructed 50 cm above ground level. Such a seedbed will create favourable moisture conditions for the plants by increasing the water supply and decreasing evaporation.

Ten-month-old seedlings of *Atriplex canescens, A. halimus, A. glauca* and *Simmondsia chinensis* from the KISR nursery were transplanted. Holes of 50 cm diameter were made along the embankments at 40–50 cm depths and 100 cm apart. Plants were removed from flower pots, placed in holes and the soil was firmly pressed around the plant roots. The numbers transplanted were: *A. halimus*, 387; *A. glauca*, 525; *A. canescens*, 60; and *S. chinensis*, 41.

Plant survival was monitored and measurements of plant height and width were recorded.

RESULTS AND DISCUSSION

1. *Seed germination experiments on native plants*

Germination tests in six simulated environments: Table 3 is a summary

Table 3. Germination percentage of native species under various simulated environmental conditions.

	Germination (%)					
Species \ *Month*	*Oct.*	*Nov.*	*Dec.*	*Mar.*	*Apr.*	*May*
Shrubs						
Zygophyllum coccineum	94	45	76	—	—	52
Convolvulus oxyphyllus	0	3	—	—	—	—
Rhanterium epapposum	0	5	—	2	—	—
Ochradenus baccatus	9	—	—	—	9	2
Heliotropium bacciferum	0	—	—	—	—	0
Perennial grasses						
Panicum turgidum	38	90	—	—	—	—
Pennisetum divisum	19	7	—	—	—	14
Aeluropus lagopoides	100	100	—	—	50	—
Annual grasses						
Hordeum glaucum	93	95	86	63	73	39
Stipa capensis	97	98	98	—	35	16
Schismus barbatus	50	53	—	—	—	33
Cutandia memphitica	77	41	69	—	10	77
Bromus madritensis	—	—	—	—	91	93
Forbs						
Bassia eriophora	2	21	5	13	39	5
Astragalus schimperi	—	36	—	—	—	36
Malva parviflora	—	—	—	—	—	3

— = not determined.

of germination results obtained from the six simulated environmental conditions. When considering all the vegetation strata, it is evident that the annual species were able to germinate better than other vegetation types. This was followed by the perennials, shrubs and forbs.

Germination data for October showed that *Aeluropes lagopoides* had the highest percentage, followed by *Stipa capensis*, *Zygophyllum coccineum*, *Hordeum glaucum*, *Ochradenus baccatus* and *Cutandia memphitica*. *Convolvulus oxyphyllus*, *Rhanterium epapposum*, and *Heliotropium bacciferum* failed to germinate in this environment. Seed viability seems to influence greatly the percentage of germination. As for November data, annual grasses again showed the highest percentage, *Aeluropus lagopoides* produced the largest number of plumules, whereas *Convolvulus oxyphyllus* had the lowest. In the December-simulated environment only a few species were tested. Of these, *Stipa capensis* showed the highest percentage germination, whereas the forb *Bassia eriophora* produced the lowest number of plumules. Annual vegetation was again dominant in the May-simulated environments.

The failure of *Rhanterium epapposum*, *Convolvulus oxyphyllus* and *Heliotropium bacciferum* to germinate suggests either that these seeds were not viable or that they require seed treatment prior to germination. Future tests of these species will include some physical, chemical or mechanical treatment of the seeds.

Annual vegetation seems to possess the ability to grow in a wide range of environments (Table 3). Species produced significant numbers of plumules in the fall, winter and spring. Additionally, the germination percentage was generally in the upper 70%, reaching as high as 100% for *Aeluropus lagopoides*. Though the production of these species is generally low and seasonal, their ability to germinate in various environments makes them an important vegetation element for future investigations. Moreover, these species were observed to provide a significant portion of livestock diet during the rainy season.

Forb germination was found to be generally low (Table 3). Species failed to produce a significant number of plumules in all simulated environments. The November-simulated environment was the only one in which more than one species showed percentage germination higher than 20%. This makes it necessary to use some type of seed treatment prior to germination for this vegetation component.

When considering the six environmental conditions, it is evident that October- and November-simulated environments were the most favourable ones for seed germination. The December-simulated environment was the least favourable one for plant germination and the May environment was half-way between the two extremes. This suggests two things: firstly, that October and November are appropriate times for seeding, and secondly, that there exists some similarity in the physiological characteristics and responses of some of the seeds, which perhaps raises the possibility of planting these species at the same time. Should further tests of these species prove conformity, similar dates for planting or revegetation of these species could then be recommended. This would be advantageous in terms of cultural practices, manpower utilization and establishment of mixed stands.

2. *Germination tests for exotic species*

In denuded rangelands where the seed source is meagre and the natural recovery of the range is slow, introduced species offer a suitable alternative for animal feed as well as being useful in revegetation trials. Before bringing such species for large-scale applications, however, it is necessary to look into many variables, such as growth, establishment, production, palatability, nutritive value and resistance to grazing pressures. The data obtained from such investigations would provide the basis for a successful species introduction program. Following is a discussion of these variables from experimental data obtained for some exotic species.

(a) Seed germination under six environmental conditions: Table 4 shows species response to the six tested simulated environments. Of the vegetation strata, the shrubby species showed a higher percentage of germination than did the other vegetation types. *Atriplex* species produced more plumules than the other plants in these environments. Environments 2 and 6 were the most favourable for this germination. As for the grasses, Environment 1 gave the highest number of plumules

Table 4. Germination percentage of exotic species under various environmental conditions.*

Species	*Environment used**					
	1	*2*	*3*	*4*	*5*	*6*
Shrubs						
Atriplex halimus	—	33	—	84	83	83
Atriplex glauca	—	95	—	—	99	—
Atriplex lentiformis	—	83	—	—	91	—
Atriplex semibaccata	—	78	—	—	59	72
Atriplex canescens	—	66	—	77	—	11
Atriplex confertifolia	—	2	—	2	—	0
Atriplex cuneata	—	1	—	—	—	—
Kochia prostrata	—	3	—	17	—	—
Artemisia tridentata	—	—	—	2	—	—
Simmondsia chinensis	92	—	—	—	—	—
Forb						
Achillea millefolia	—	—	—	—	—	100
Grasses						
Cenchrus ciliaris	—	—	—	17	24	45
Pennisetum setaceum	—	—	—	14	5	—
Pennisetum divisum	—	—	—	50	57	—
Agropyron cristatum	10	96	—	26	10	—
Agropyrum trachycaulum	1	—	—	5	2	—
Agropyron desertorum	14	—	—	24	10	—
Agropyrum riparium	100	—	—	—	—	—
Dactylis glomerata	100	96	97	—	—	—
Sporobolus airoides	100	—	99	—	—	—
Eragrostis curvula	100	—	—	—	—	—
Avena orientalis	100	—	—	—	—	—
Cynodon dactylon	—	—	—	—	—	2
Bragutaria decumbens	—	—	—	—	—	25

*See Table 2 for description of environments.
— = not determined.

(Table 4); five species showed 100% germination. Environment 3 was also favourable as the species being tested showed 96–99% germination. Grass species tested in environment 6 produced the smallest number of plumules.

When considering all the environments, it is evident that Environments 1 and 2 were the most favourable for seed germination. Both shrubs and forbs showed higher percentage germination under these environments. This suggests the possibility of planting those species at the same time, which is advantageous in terms of cultural practices and manpower utilization.

(b) Seed germination under Kuwait's simulated conditions: Table 5 summarizes data for variation in percentage germination for the four *Atriplex* species under each of the four simulated environments. Percentage germination for each species did not differ significantly for

Table 5. Percentage germination of four exotic species in four different environments.

Treatments (Environments)	Atriplex glauca	A. halimus	A. canescens *(19 FUS)*	A. canescens *(20 SYAG)*	*Mean*
February	49.0a	23.3a	10.9a	56.2a	38.8a
March	38.0b	9.5b	14.2a	22.8b	21.2b
April	31.4b	21.4a	1.9b	10.5c	16.3c
October	35.4b	20.9a	3.8b	23.3b	20.9bc

Means followed by the same letter in the same column are not statistically significant at the 0.05 level according to the LSD test.

March, April and October conditions. This indicates that the February environment is, perhaps, the most suitable. During this period, the plants experienced environmental conditions of 20.7 and 9.3°C by day and night, respectively. Relative humidity varied between 31 and 80% for the same period. *A. halimus* gave the same number of plumules in all treatments except for March, which was significantly lower. It should be noted that germination percentage for all treatments (9.3–23.3) is generally low, presumably because seeds had not been removed from the utricles.

Data for *A. canescens* (19 FUS) are presented in Table 5. Highest germination was found in seeds germinated in February and March. The number of plumules in all treatments was small and varied between 10.9 and 1.9%. Apparently, most of the seeds used were still retained in the utricles, which inhibited germination. On the other hand, seeds of *A. canescens* (variety 20 SYAG) showed relatively good germination results. The number of plumules produced under February conditions was significantly higher than in other environments. The good germination results for this species indicate that the species is able to germinate fairly well without the removal of the utricle and, therefore, is a potential species for revegetation without germination enhancement or other seed treatment.

Data in Table 5 indicate that the four tested species seem to react similarly to the various environments. Plants germinating under February conditions generally had the highest number of plumules in relation to other environments. *A. canescens* (20 SYAG) and *A. glauca* had the highest germination percentages, *A. halimus* and *A. canescens* (19 FUS) were inferior.

Examination of the response of *Atriplex* spp. to other environments (March, April and October) shows a trend similar to the February environment. Compared to the other two species, *A. glauca* and *A. canescens* (20 SYAG) were found superior in each environment except April. Their superior performance in each environment was attributed partly to the inherent ability of the seeds to germinate without removal of the fruits. The most significant factor, however, is the selection of the environment in which most species germinate best rather than individual variations within each environment. This knowledge is useful when selecting the planting time of mixed stands in range revegetation.

Based on the data presented in Table 5, it is evident that the February environment is the most suitable for the germination of these *Atriplex* spp. in Kuwait. These findings are in line with studies reported by Springfield (1970) in which *Atriplex* spp. germinated best at an average daily temperature of 20.5°C. Similar results were reported by Beadle (1962). Plants grown under October and April environments did not show marked differences. Due to the harsh environmental conditions encountered in the summer (June–September), however, it is not advisable to sow these species close to the onset of summer (April), so the October planting would be advantageous in such cases.

(c) Seed germination in two soil media: In areas where climate and soil impose limitations on production of local vegetation, the intoduction of new species suitable for the environment is often advanced (Hartley 1963, Stubben and McCulley 1977). Testing and evaluation of the plant species for areas that are climatically and edaphically analogous to Kuwait was expanded to include use of available agricultural soil.

Percentage germination of the species is shown in Table 6. Data indicate that, in general, soil type had no effect on germination. The

Table 6. Effect of soil media on percentage germination of six Atriplex species (Environment no. 2).*

Species \ *Soil media*	*Peatmoss + agr. soil*	*Peatmoss only*
Atriplex canescens	84.0a	81.7a
Atriplex halimus	71.4a	61.1a
Atriplex inflata	88.6a	82.8a
Atriplex glauca	92.9a	85.8a
Atriplex lentiformis	5.7a	19.1b
Atriplex confertifolia	0	0

a, b: along the same row indicate significance at $P<0.10$ level of probability according to the LSD test.
*See Table 2.

failure of *A. confertifolia* to germinate may be due to low seed viability. *A. lentiformis* appears to germinate better in 100% peatmoss soil, but this point should be investigated further. It appears that soil type did not affect germination percentage. This conclusion is of importance in terms of material cost as well as availability. Peatmoss is a costly material and its availability is seasonal, which makes the agricultural soil mix a much more attractive alternative. Should further tests confirm this, it is recommended that the agricultural soil mix rather than peatmoss be used for germination studies of these shrubby species. Findings from both experiments concerning germination behaviour of these and other shrubby species will be tested further under different environmental conditions using different soil media before final recommendations are made. It is also hoped that other shrub species will be included in future experiments.

3. *Establishment of desert halophytes*
   (a) *Use of saline waters*
      (i) *Experiment I – Atriplex halimus:* The effect of various treatments on the height of *Atriplex halimus* during the 20 week experiment is shown in Table 7; width measurements are shown in Table 8. In discussing the results in these tables, the data should be interpreted with reference to the water composition of each irrigation treatment (Table 9). The predominant soluble ions in brackish water, sewage and seawater varied considerably in amount. Of particular interest are sodium, potassium and chloride which showed great variability in the nine water treatments.

When considering the height of *Atriplex* seedlings treated with nine different irrigation waters, it becomes evident that variation occurred between treatments and with time (Table 7). Significant differences between treatments occurred within the third week when sewage water gave the best response (14.4 cm) compared to the seawater dilution of 1:7 (6.8 cm). By the ninth week, plants irrigated with brackish and sewage water gave best response, with no significant differences between them. The same relation persisted up to the 20th week. From the 17th week onwards, seawater dilution of 1:15 (2,500 ppm) proved superior to other dilutions in terms of positive plant growth responses. Dilutions of 1:3 and 1:1 (10,000 and 20,000 ppm) were also found to allow plants to grow reasonably well through the 20th week. Weaker dilutions (3:1 and 1:0 sea = fresh water) were found to affect plant height adversely.

Data for plant width and irrigation waters are shown in Table 8. Plant width showed a response to treatments similar to that of plant height.

From the preliminary data obtained on the response of plant height and width to irrigation treatments, it could be concluded that brackish water, sewage and seawater dilutions of 1:15, 1:7, 1:3 and 1:1 could be used for seedling establishment of the *Atriplex* species. Since the time for seedling establishment is prior to transplanting, the results suggest the possibility of using these available water sources for large-scale planting of *Atriplex*. More research is needed, however, before data become conclusive.

(ii) *Experiment II – Atriplex halimus, A. glauca and A. canescens:* The present experiment was conducted as a follow-up to a previous study (Nassef *et al.* 1980) which indicated that the best plant growth response of *Atriplex halimus* was achieved under

Table 7. Effect of nine water treatments on height (cm) of *Atriplex halimus* seedlings.

Treatments / *Weeks*	*3*	*5*	*7*	*9*	*11*	*13*	*15*	*17*	*19*	*20*
Sewage	14.4a	18.7a	21.1a	31.6a	33.8a	37.9a	53.3a	52.9a	49.5a	53.8ac
Brackish	9.8c	13.5ab	20.8a	34.8a	40.1a	43.7a	57.4a	61.9a	61.4a	64.5a
Seawater dilutions										
1:15	8.9bc	12.0ab	14.7c	20.4c	23.8c	28.4c	44.9c	46.1c	47.0c	52.5c
1:7	6.8b	8.5b	9.8bd	12.2bd	15.7de	18.7de	26.6c	29.7d	30.7d	32.5d
1:3	8.2bc	9.5b	11.3bcd	15.3cd	18.2cd	21.7cd	23.7cd	25.0d	26.6de	26.7d
1:1	9.4c	11.1b	12.3cd	14.5d	15.0de	16.1de	17.1cd	18.6de	19.2ef	26.7d
3:1	8.8bc	9.9b	13.0cd	11.6b	11.9e	12.0e	13.9cd	12.5e	11.3bf	13.1e
1:0	8.7bc	9.0b	8.2b	7.3b	4.4b	0.8b	0.8b	0.8b	0.9b	0.9b
Control	9.7c	8.1b	9.8bd	10.1bd	10.8bde	12.1e	12.8bd	13.0e	12.5f	14.8e

a, b, c, d, e, and f: means with the same letter within the same column are not significantly different at the 0.05 level of probability (LSD).

Table 8. Effect of nine water treatments on crown width (cm) of *Atriplex halimus* seedlings.

Treatments / *Weeks*	*3*	*5*	*7*	*9*	*11*	*13*	*15*	*17*	*19*	*20*
Sewage	6.8a	6.8a	7.1c	16.7c	25.8a	36.5a	47.2c	54.3c	50.7a	50.3a
Brackish	5.6c	6.8a	9.8a	26.6a	33.4a	47.4a	64.7a	63.3a	52.2a	59.5a
Seawater dilutions										
1:15	5.5c	6.3ac	7.2c	9.3d	11.4c	13.0c	24.5d	35.3d	36.3c	32.0c
1:7	4.1b	4.4bdc	4.8bd	5.3bd	5.2bc	8.5bc	11.9be	16.3b	15.5d	17.3d
1:3	4.3b	4.9dc	5.3d	5.0bd	6.7bc	6.3bc	7.8bc	14.8b	11.8bd	11.6bd
1:1	4.4bc	5.3bc	5.1bd	5.3bd	5.9bc	5.6bc	5.5be	6.3b	5.3bd	14.2bd
3:1	6.0ac	5.8adc	5.0bd	5.6bd	5.5bc	4.7bc	4.5be	4.2b	4.5b	5.0bd
1:0	5.6c	5.1dc	4.2bd	3.3b	1.9b	0.5b	0.4be	0.5b	0.5b	0.5b
Control	4.0b	3.3b	3.4b	3.2b	3.4bc	3.4bc	3.2be	3.5b	3.2bd	3.1bd

a, b, c, d, e, and f: means with the same letter within the same column are not significantly different at the 0.05 level of probability (LSD).

Table 9. Analysis of the nine water types used in irrigation.

Parameter	Sewage (ppm)	Brackish (ppm)	Seawater dilutions 1:15 (ppm)	1:7 (ppm)	1:3 (ppm)	1:1 (ppm)	3:1 (ppm)	1:0 (ppm)	Control (potable (ppm) water)
Na^+	340	540	720	1300	3000	5700	7800	11700	55
Ca^{2+}	150	397	30	59	116	232	365	465	36
Mg^{2+}	63	157	92	179	362	722	1146	1493	27
K^+	23	14	29	55	175	210	340	540	2
Cl^-	404	815	1397	2748	5672	10813	17318	22043	81
SO_4^{2-}	450	1600	185	360	750	1550	2500	3100	185
CO_3^{2-}	8	9	N.D.	N.D.	5	11	15	21	6
HCO_3^-	349	139	11	20	32	66	102	134	10
PO_4^{3-}	14	N.D.	N.D.	0.04	0.04	0.04	0.08	0.06	0.02
NO_3^-	148	2.69	N.D.	N.D.	N.D.	N.D.	N.D.	0.09	0.07
NO_2^-	1.460	0.008	N.D.	N.D.	N.D.	N.D.	0.008	0.008	0.030
SiO_2	11.78	15.7	N.D.	0.07	0.35	0.66	1.02	1.33	2.12
pH	8.1	8.0	7.5	7.7	8.0	8.0	8.0	8.0	7.4
TDS (g/l)	1.87	3.64	2.68	5.03	11.20	21.30	33.80	44.96	0.38

Analysis done at the Central Analytical Laboratory, KISR; N.D. = Not detectable.

seawater dilutions between 1:7 and 1:15. Hence, the overall objectives were to use seawater dilutions between 1:7 (5,000 mg l^{-1}) and 1:15 (2,500 mg l^{-1}) to find the optimal level. Dilution levels of 1:9 (4,000 mg l^{-1}), 1:11 (3,333 mg l^{-1}) and 1:13 (2,857 mg l^{-1}) were used; control was 0:1 (fresh water).

The effects of irrigation treatments on the three *Atriplex* spp. are presented in Table 10. The height of the four-week seedlings was measured at the start of the experiment and every two weeks afterwards through to the 12th week. The growth response of *A. canescens* depended on the treatment. Plants irrigated with a seawater dilution of 1:7 showed the best response by the 8th week and continued to respond well through to the end of the experiment (33.2 cm). *A. canescens* grown under other seawater dilutions (1:9, 1:11, 1:13, 1:15 and control) did not exhibit any marked variation. Differences in height for these treatments were generally not significant from the 4th week until the end of the experiment. At the 12th week height varied between 21.9 and 25.6 cm for the 1:15 treatment and for the control, respectively. This suggests that *A. canescens* adjusts well to the irrigation water within the salinity range of these treatments. It further indicates that the plant grows better under a higher salinity, 1:7 (5,000 mg l^{-1}) than under a more diluted seawater of 1:15 (2,500 mg l^{-1}). The ability of the plant to thrive at a higher salinity is attributed to its ability to store Na^+ and balance it by the synthesis of the oxalate anion in the leaf surface (Mozafer and Goodin 1970). Whenever the concentration of salt reaches a lethal level, it seems that the balloon-like cells burst, relieving the plant of excess salt. It should also be noted that the nutrients supplied by the seawater, such as PO_4^{3-}, NO_3^- and K^+, and the ability of the plant to withstand saline water were probably the prime reasons for good plant performance under these treatments.

Atriplex halimus showed a somewhat different trend from that exhibited by *A. canescens* (Table 10). Measurements of plant height showed that the plant generally did not vary significantly in growth among the treatments. Apparently, the plant's tolerance level was not exceeded between 2,500 mg l^{-1} (1:15) and 5,000 mg l^{-1} (1:7). It was observed that the plant had performed better in the control for the first 10 weeks of the experiment and differed significantly from all treatments except in the 12th week, where it was similar to the 1:13 and 1:9 treatments. This, perhaps, suggests a lower salinity tolerance than *A. canescens*. However, it should be mentioned here that the plant growth in the control did not exceed by 50% most of those under saline treatment.

Measurement of plant height for *A. glauca* showed a similar response to that of *A. halimus* under the saline treatments. Although plants grown under the 1:7 treatment were generally

Table 10. Effect of different salinity levels on heights of three *Atriplex* spp.

Species	Treatment (seawater: fresh water ratio)	Height (cm)						
		0	*2*	*4*	*6*	*8*	*10*	*12 Weeks*
Atriplex canescens	1:7	7.2b	14.1a	19.6a	21.7a	24.8a	28.3a	33.2a
Atriplex canescens	1:9	6.4b	12.0b	15.7bc	18.5abc	19.6b	21.4b	23.3b
Atriplex canescens	1:11	9.1a	12.3ab	14.6c	16.5bc	17.5b	19.9b	22.8b
Atriplex canescens	1:13	7.7b	11.6b	13.8c	15.3c	16.5b	19.6b	24.5b
Atriplex canescens	1:15	7.6b	13.0ab	17.7ab	19.5ab	19.6b	20.7b	21.9b
Atriplex canescens	control	7.0b	12.5ab	15.9bc	16.7bc	19.3b	23.0b	25.6b
Atriplex halimus	1:7	4.5c	9.5b	10.0c	15.4c	19.3cd	24.2bcd	27.0bc
Atriplex halimus	1:9	6.1b	11.7a	16.1a	21.8ab	25.0b	27.1bc	30.0ab
Atriplex halimus	1:11	8.0a	11.4ab	12.2bc	15.9c	17.9d	20.1cd	22.7bc
Atriplex halimus	1:13	7.5a	11.4ab	14.9ab	20.0b	24.3bc	30.0ab	30.4ab
Atriplex halimus	1:15	5.8b	10.2ab	11.5c	14.1c	16.0d	17.6d	19.1c
Atriplex halimus	control	5.9b	12.0a	17.8a	25.7a	32.7a	36.5a	38.0a
Atriplex glauca	1:7	8.4ab	15.4a	21.0a	25.3a	30.0a	32.6a	36.6a
Atriplex glauca	1:9	6.8b	15.2a	14.9b	20.1b	22.6b	26.6b	29.4a
Atriplex glauca	1:11	9.7a	15.2a	19.2a	21.4ab	23.2b	24.5a	26.4b
Atriplex glauca	1:13	8.1ab	15.7a	21.5a	23.9ab	26.8ab	30.4a	33.3ab
Atriplex glauca	1:15	8.0ab	15.2a	19.3a	19.8b	23.0b	25.7a	27.4ab
Atriplex glauca	control	8.4ab	15.0a	19.2a	23.3ab	25.8ab	31.0a	32.8ab

Means followed by the same letter within each column for the same species are not statistically significant.

superior to those grown under other treatments, the differences were not always significant (Table 10). Similarly, the response of plants under other irrigation treatments, as well as in the control, did not differ significantly.

From Table 10, it can be seen that the species tested tended to favour the higher rather than the lower salt concentrations. Thus, difference in dilution levels between 1:7 and 1:15 was of no great significance in seedling establishment of these three *Atriplex* spp. This result suggests the possible use of available saline water of 5,000 $mg\,l^{-1}$ (total dissolved solids). It also indicates the possibility of planting the three species together as their tolerance levels are comparable. This would be advantageous in terms of establishing mixed stands, which are usually preferable for revegetation and grazing purposes.

Another interesting application of the results of this research is in the area of shrub transplanting. As shrub seedlings generally require 12–15 weeks of irrigation prior to transplanting, it can be assumed that such seedlings can be irrigated with Kuwait's available saline water of 5,000 $mg\,l^{-1}$ while the plants are in nurseries or until the appropriate time for transplanting in the field.

(b) *Dry farming practices*

Plant response to dry farming practices at different times is shown in Table 11. Evaluation of *A. glauca*, *A. canescens* and *Simmondsia chinensis* was conducted at eight and 43 weeks following treatment. *A. halimus* was investigated at 15 and 43 weeks after transplanting.

Initial evaluation of the species eight weeks after transplanting was very encouraging. All plants showed high survival percentage. Similarly, plant production parameters (height and width) gave good results.

With the onset of summer in 1981, the growth of all plants was observed to slow down and they became dormant. By mid-July, little if any growth was apparent. Despite the harsh environmental

Table 11. Evaluation of production parameters and percentage survival for shrub species under dry farming.

	8 Weeks			*15 Weeks*			*43 Weeks*		
Species	*Survival (%)*	*Height (cm)*	*Width (cm)*	*Survival (%)*	*Height (cm)*	*Width (cm)*	*Survival (%)*	*Height (cm)*	*Width (cm)*
Atriplex halimus	—	—	—	95.3	118.1	45.2	8.9	91.6	63.8
Atriplex glauca	96.4	65.1	73.8	—	—	—	55.9	29.0	47.0
Atriplex canescens	93.3	65.8	73.5	—	—	—	15.0	46.3	40.3
Simmondsia chinensis	100.0	—	—	—	—	—	7.3	22.0	12.0

summer conditions and the exceptional temperatures recorded during the season, no irrigation was carried out. Following the summer (43 weeks after transplanting), an evaluation of plant response under dry farming was conducted (Table 12). It was evident that the high temperature extremes and solar radiation had adversely affected the plants. Percentage survival of *A. halimus* dropped markedly from 95.3 to 8.9%; *A. glauca* declined from 96.3 to 55.9%; *A. canescens* was also severely affected and decreased to 15.0%, whereas *Simmondsia chinensis* declined to 7.3%.

Though percentage survival was low for all species except *A. glauca*, it should be mentioned that the delay in the onset of the rainy season of 1981–1982 also contributed to the low survival values. Based on these limited data, it may be concluded that *A. glauca* seems to be a good candidate for dry farming practices but the shrub has the inherent disadvantages of being short and of low biomass productivity. The results reported are preliminary and hence further studies are needed before final recommendations can be made.

SUMMARY AND CONCLUSIONS

Seed germination experiments were conducted for both native an exotic species. Seeds were collected and tested from 16 species which included grasses, forbs and shrubs. They were tested under simulated Kuwait environments based on a 25-year average of climatological data for October, November, December, March, April and May. Data showed that annual grass species were easier to germinate than forbs, perennial grasses and shrubs. It also revealed that species differed greatly in seed viability, as well as optimum environmental conditions for germination. The experiments furnished valuable information to determine appropriate planting dates for these native species as well as shedding light on species' response to various cultural practices.

Germination tests were also conducted on 24 exotic species. These were tested with large-scale planting in the depleted ranges of Kuwait in view. Plants were germinated in simulated and natural environments under Kuwait climatological conditions of October, November, February, March, April and May. Some species were promising, e.g. the grasses *Cenchrus ciliaris*, *Pennisetum setaceum*, *Eragrostis curvula*, *Sporobolus airoides* and *Agropyron riparium*, and the shrubs *Atriplex canescens*, *A. halimus*, *A. glauca* and *A. lentiformis*.

The native and exotic species tested exhibited varied germination responses under the different climates, with *Bromus madritensis*, *Hordeum glaucum* and *Stipa capensis* showing the best results and more environmental tolerance. Experimentation with *Atriplex* species showed that *A. canescens* gave the best response and had a better ability to germinate under varied environmental conditions.

Use of various soil media showed the suitability of Kuwait soil for germination trials rather than using pure foreign soil media (vermiculite and peatmoss), which is currently practised. Further tests are needed under range conditions before recommendations are made.

Irrigation waters available in Kuwait were investigated for possible use in establishment of desert halophyte seedlings. Sewage, brackish, diluted and undiluted seawater were all tested. Preliminary data showed the possibility of using these waters for the *Atriplex* species. Seedlings grew without adverse effects for a 20-week study period. Studies will be continued for long-term effects. The *Atriplex* species responded better than the jojoba plant (*Simmondsia chinensis*) to dry farming experiments. Evaluation after field transplantation and non-irrigation of the plants showed that only *Atriplex glauca* had reasonable survival. The harsh summer conditions of 1981 markedly affected the tall growing shrub species (*A. halimus* and *A. canescens*). Monitoring of the environment's long-term effects on plant establishment and productivity will be continued.

LITERATURE CITED

Beadle, N. C. 1962. Studies in halophytes. The germination of the seed and establishment of the seedlings of five species of *Atriplex* in Australia. *Ecology* **33:** 49–62.

Behbehani, H., F. K. Taha and A. A. Nassef. 1981. A preliminary germination study of *Atriplex* species under Kuwait's environment. Annual Research Report, Kuwait Institute for Scientific Research (1981): 17–19.

Chatterton, N., J. R. Goodin, C. M. McKell, R. V. Parker and J. M. Rible. 1971. Monthly variation in chemical composition of desert saltbush. *J. Range Manage.* **24:** 37–40.

Dowling, P. M., R. J. Clements and J. R. McWilliam. 1971. Establishment and survival of a pasture species from seeds sown on the soil surface. *Austr. J. Agric.* **21:** 61–74.

Gale, J., R. Noaman and A. Mayber. 1970. Growth of *Atriplex halimus* in sodium chloride salinated culture solution as affected by relative humidity of the air. *Austr. J. Biol.* **23:** 947–952.

Goebel, C. J. 1978. Grass establishment on Lower Snake River Rangelands. Proc. I Int. Rangeland Cong., Denver, Colorado, pp. 629–630.

Goodin, J. R. and C. M. McKell. 1970. *Atriplex* spp. as a potential forage crop in marginal agricultural areas. In Proc. XI Int. Grasslands Cong., Univ. Queensland Press, pp. 158–161.

Hartley, W. 1963. The phytogeographical basis of pasture plant introduction. *J. Genetica Agraria* **17:** 1–4.

Kernick, M. D. 1964. Plant resources and range ecology and fodder plant introduction. FAO Report No. TA2181 to the Government of Kuwait, Rome, Italy.

Knowles, G. H. and R. W. Condon. 1951. The perennial saltbush (*Atriplex vesicarium*). *J. Soil Conserv. Serv. (New South Wales)* **7:** 1–9.

McGinnies, W. J. 1960. Effect of moisture stress and temperature on germination of six range grasses. *Agronomy J.* **52:** 159–162.

Mott, J. J. 1977. Factors affecting seed germination in three annual species from an arid region of Western Australia. *J. Ecol.* **62** (3): 399–709.

Mozafer, A. and J. R. Goodin. 1970. Vesiculated hairs. A mechanism for salt tolerance in *Atriplex halimus*. *Plant Physiol.* **45:** 62–65.

Nassef, A. A., F. K. Taha and S. A. Omar. 1981. Effect of defoliation and irrigation water on growth and production of *Atriplex glauca.* Annual Research Report, Kuwait Institute for Scientific Research (1980): 15–17.

Nemati, Nasser. 1977. Comparative palatability of *Atriplex canescens*. *J. Range Manage.* **30:** 268–369.

Omar, S. A. and F. K. Taha. 1981. Effect of environmental factors and soil media on seed germination of forage species. Annual Research Report, Kuwait Institute for Scientific Research (1980): 13–15.

Plummer, A. P. and G. T. M. Fenley. 1950. Seasonal periods for planting grasses in the sub-alpine zone of central Utah. *Intermountain Forest and Range Exp. Sta.* Res. Paper 18, Ogaden, Utah.

Plummer, A. P., S. B. Perry, D. R. Monsen and D. R. Christensen. 1966. Four-wing saltbush, a shrub for future game ranges. Pub. No. 66, Utah St. Dep. Fish. and Game, Salt Lake City, Utah.

Springfield, H. W. 1970. Germination and establishment of four-wing saltbush in the Southwest. USDA Serv. Res. Paper No. RM-55. Washington D.C., USA.

Stubben, D. J. and G. Wayne McCulley. 1977. Effect of temperature and photoperiod on germination and survival of sand bluestem. *J. Range Manage.* **29** (3): 206–208.

Taha, F. K. and S. A. Omar. 1980. Preliminary study of seed germination of eight native plant species from Kuwait rangelands. Annual Research Report, Kuwait Institute for Scientific Research (1979): 5–6.

Taha, F. K., S. A. Omar and I. M. Bilal. 1983. Range management model: An integrated system for resource evaluation, forage production and de-desertification in Kuwait. Project's Final Report. Kuwait Institute for Scientific Research, Techn. Rep. No. 969.

Taha, F. K., M. A. Razzaque and S. A. Omar. 1984a. Range Management Phase II. Assessment and improvement of rangelands of Kuwait for forage and animal production without desertification. Proposal accepted by KISR. Publication No. 1327 B, May 1984.

Taha, F. K., M. A. Razzaque and S. A. Omar. 1984b. Range Management Project Phase II. A proposal accepted by Kuwait Institute for Scientific Research.

Biomass and forage production in the arid and semi-arid zone

J. R. GOODIN

Department of Biological Sciences, Texas Tech University, Lubbock, Texas 79409, USA

ABSTRACT

One need only look at the existing vegetation in the arid and semi-arid zone to realize that there are ecological and genetic reasons to be optimistic. Few deserts are bare, and in most of those cases there are anthropogenic reasons for lack of vegetation. One must be careful not to equate *quality* extant biomass with potential productivity in the arid zone. Many arid regions not now producing measurable forage may do so if the grazing pressures are controlled and if the proper germplasm can be established. Exclosures have demonstrated in many regions with 50–200 mm precipitation that desirable plants can be established, and that they will survive for long periods of time.

There is growing evidence that perennial shrubs provide a far better opportunity for utilizing the edaphic and environmental resources of an arid region than either annual or perennial grasses. Although ephemeral grasses and forbs exhibit an amazing potential for utilizing limited and unpredictable rainfall, their temporary nature subjects the landscape to erosion for much of the year, increased albedo from the soil surface, and, in years of extreme drought, no cover whatsoever. Perennial shrubs, on the other hand, lend an ecological stability to the landscape and provide at least some cover during extreme drought, although productivity may be low.

Experience in the southwestern United States and in the Middle East suggests that a number of genera within the Chenopodiaceae have evolved under a climatic regimen of exteme aridity and salinity, while producing a high protein forage. Varying degrees of palatability have subjected these plants to differing herbivore pressure, and in some cases local or total extinction has resulted for the more palatable ones. In those cases where the germplasm still exists, there is tremendous natural variability which allows for selection for site specificity. The genus *Atriplex* can be used as a model to predict the forage potential in regions where environmental and edaphic stess can be compensated with managerial skills to increase the biological productivity. New biotechnologies are becoming available which promise to help the arid zone ecologist accomplish these objectives.

INTRODUCTION

Agriculturists have traditionally dealt with the most cost-effective programs first, moving on to problem solving in difficult sites only when population pressures and changing economic patterns dictate that one should do so. Even in the semi-arid regions where environmental constraints make sustained productivity uncertain, range managers have relied chiefly on annual and perennial grasses to produce the biomass necessary for domestic livestock. On the fringes of the zones of predictable rainfall, dryland farming gives way to pastoralism, and the emphasis has been placed on maintaining and improving existing vegetation. In some sites, exotic species have been introduced successfullly, but the rate of success is rather low. Direct seeding of grasses is sometimes successful if seed-bed preparation is adequate, if precipitation occurs, and if the subsequent weather conditions favour the establishment of a tiny germinating seed lying at or near the soil surface. Years of trials with rangeland drills, aerial seeding, pelleting of seed with various materials and fertilizers, and other creative efforts have failed to provide foolproof methodology for establishment under these harsh conditions.

As one moves to the truly arid zone, the magnitude of the problem increases enormously. Not only are the cost-benefits more difficult to achieve, but the biological parameters of establishment may seem insurmountable. The traditional outcome of these problems is to give up, or conclude that the energy inputs can never justify the productivity achieved. Making the desert bloom would appear to be too costly.

ESTABLISHMENT

The problems of vegetation establishment in the arid zone are totally unlike those anywhere else in the world. Lack of precipitation, temperature extremes, high winds, and an incredible rate of evapotranspiration combine to limit establishment, growth and development only to those plants with exceptional adaptations to environmental stresses. Combine these problems with a lack of management of herbivores, and the list of possible species for range improvement grows very short.

For many years range scientists have attempted to directly seed grasses, forbs, and occasionally, shrubs. The history of these projects is largely one of failure, particularly in the drier zones. One only need look at the climatic pattern and age classes of established perennial vegetation to realize that production of perennial plants comes only on rare occasions in the desert. It is not unusual to find stands of desert shrubs of only two or three age classes in 50-year-old stands (Goodin and Newton 1981). This lack of establishment would suggest that the exact climatic conditions for not only germination, but establishment, may come along only once in many, many years. No doubt germination occurs often, but seedlings lying on the hot, parched surface of a desert floor may have little chance of developing a radicle quickly enough and deeply enough to have a fair chance of establishment. In deep desert sands, with poor water holding capacity, the chances are greatly diminished. Ephemeral vegetation, on the other hand, may grow so quickly that they take advantage of even small quantities of precipitation.

The key to successful establishent in desert situations is timing. There may be a time "window" in which seed germination and establishment can be achieved. This time may vary greatly from one region to another, and will depend on the species in question and the ecological strategy. In a region with predictable spring rains, the seeding might be done in early to mid-spring. For West Texas, predictable rains do occur in late spring, but hot, drying winds occur at the same time and one may take an inordinate risk by seeding at that time of the year. There is reason to believe that fall planting, when winds are calmer and evapotranspiration rates are lower because of the onset of cool weather, provides a better chance of survival of arid zone shrubs and perennial grasses, provided the species in question can survive the cold winters in the seedling stage.

Recent experience with transplanting of shrubs to arid sites in West Texas suggests that with the addition of only a small amount of water, it is possible to assure survival of shrubs, even under the most extreme conditions. Cost analysis of transplanting desert shrubs would appear to be favourable, especially if survival can be guaranteed, and similar studies in Libya and Tunisia confirm these results. Our own experience suggests that once the transplant has become established, there is an excellent chance for survival, provided overgrazing does not occur (Goodin and Newton 1981).

FORAGE PRODUCTION

Ecological observations in the field demonstrate that not all plants of any single species may be palatable to livestock. It is likely that no single aspect of range management is as poorly understood as is the subject of palatability. Genetic variability between plants of the same species must dictate a particular combination of metabolites that integrate to produce a particular flavour for herbivores. Some of these molecules, even in small quantities, may be distasteful to grazing animals, and therefore the forage will not be grazed, or will be grazed only as a last resort. Most animals have learned, through trial and error, that certain plants are unacceptable and/or poisonous. Various organic acids, alkaloids, phenolics, and other metabolites affect the palatability of various species and if consumed in certain concentrations, may cause mammalian toxicity.

Prior knowledge of these principles may help the scientist to select those particular plants which, for whatever biochemical reason, provide the most palatable forage. In some places in the Middle East, *Atriplex halimus* is said to be totally unpalatable to cattle and sheep. Yet in other regions, it is the preferred arid zone forage. The difference would appear to be in the genetic variability of variants within the species. At the present time, a great deal of selection is going on in this and other arid zone shrubs to test for palatability, digestibility, and weight gain.

For some heretofore unexplained reason, many desert shrubs are both rich in protein and highly nutritious. Digestibility studies of various *Atriplex* species suggest that the quality of forage is similar to domesticated high-protein forages, and the protein levels are equally high. Even in nitrogen-poor desert soils, many of these shrubs manage to assimilate nitrogen, presumably from non-symbiotic microorganisms, and incorporate it into protein. In some regions of high soil or water salinity, the ash content of forage may be quite high, but there is little evidence that these high ash levels cause any more problem than increased water intake.

Many ecologists have discounted the arid lands as having such low productivity that there is little hope of achieving economic returns. Estimates of the potential productivity for arid regions must consider natural ecosystems. The problems encountered in measuring net primary productivity in a complex ecosystem such as a forest become considerably greater in an arid zone, although fewer plant species and simpler organization offer a special opportunity for comprehension of bioenergetics. Massive sampling is required because of the heterogeneity of form, life cycles, and microclimate. Ideally, each individual of each species present should be sampled to get a true picture. Also, sampling should be carried on throughout the year to assess natural abscission of plant parts, as well as primary production consumed by animals.

Under desert conditions with water as a limiting factor in growth and development at any point in time, such a task would prove almost impossible. Desert annuals, for example, may contribute greatly to net primary productivity, but due to their ephemeral nature, time of use is extremely critical. Even perennial shrubs may lose their leaves in time of severe water stress and then replace them quickly as soon as precipitation occurs.

Specific studies of productivity in non-arid regions are well documented, but few studies have been performed in arid regions. A "moist southern desert" in the Soviet Union produced 1,200 kg/ha (Lavrenko *et al.* 1955), and ephemeral herbs in Nevada produced only 400 kg/ha (Odum and Odum 1959).

The primary productivity of the desert shrub *Larrea divaricata* (creosote bush) has been studied in detail by Chew and Chew (1965). Their data show a total productivity of 1,000 kg/ha for a relatively young shrub community. They also found that plant density varied inversely with the age of the plant. Thus, as the stand aged, density decreased from 4,460 plants/ha to 2,700 plants/ha 10 years later. Biomass production peaked at approximately 1,500 kg/ha in 20-year-old plants, and productivity was back to 1,000 kg/ha/yr in a near-climax community.

As one attempts to extrapolate forage productivity from natural ecosystems, the complicating factor of grazing and partial top removal must be superimposed on the estimates. Management of grazing systems can be most beneficial in maximizing the biomass return for shrublands.

Recent studies in West Texas and elsewhere suggest that semi-domesticated and managed stands of transplanted *Atriplex canescens* (four-wing saltbush) are capable of producing 5,000 kg/ha/yr in the 250–400 mm precipitation zone (Goodin and Newton 1981). These yields are comparable to those being achieved for short rotation hardwood species under high rainfall conditions in the eastern United States. It appears that the ideal climate for *Atriplex* is one with long days, cloudy weather and moderate rainfall.

THE CASE FOR *ATRIPLEX*

The genus *Atriplex* is a cosmopolitan group of some 200 species. The group is botanically interesting, particularly so because of the adaptation to many diverse, and generally harsh, environments (Kelley *et al.* 1981). It is becoming apparent that many of the habitats of the species (rangeland, desert, salt marsh, etc.) are becoming increasingly important as a valuable natural resource.

The halophytic species of the genus are usually considered facultative (Waisel 1972) rather than obligate halophytes. Such species accumulate salts in unique trichomes on the leaf surfaces. These trichomes may eventually burst, spreading wall debris and salt crystals on the surface of the leaf until precipitation washes the soluble salts back to the soil surface for recycling (Mozafar and Goodin 1970). The suggestion has been made that these halophytes have been restricted to saline areas because of their inability to compete with glycophytes under non-saline conditions (Chapman 1975).

Many *Atriplex* species assimilate carbon via the C_4 pathway of photosynthesis and have corresponding anatomical characteristics (Laetsch 1974). Although both C_3 and C_4 species appear in the genus, it would appear that under desert conditions, the C_4 species have a distinct advantage. Most of the dominant and important *Atriplex* species, particularly the perennial ones, have the C_4 pathway.

The perennial, shrubby nature of most saltbush species is of particular significance. They often dominate the landscape and provide high protein forage for wildlife and domestic livestock (McKell 1976). The high water use efficiency, palatability, and high protein levels make several *Atriplex* species the vegetation of choice for many arid zone regions. Although certain other genera in the Chenopodiaceae, Amaranthaceae, and Aizoiaceae contribute to forage production in desert regions, the world literature would suggest that *Atriplex* has the greatest opportunity for improving rangelands with marginal precipitation.

A number of important saltbush species, including *A. canescens*, *A. halimus*, *A. nummularia*, and *A. polycarpa*, have exceptional genetic diversity. Some species, such as *A. canescens*, are spread over vast climatic and edaphic conditions. Four-wing saltbush grows from Canada, through the western United States, and into the Sonoran and Chihuahuan Deserts of Mexico. It occurs in many soil types and at all elevations. It grows under both saline and non-saline conditions.

One need only look at the domestication process of many of our crop plants to realize that homozygosity occurs only after long periods of time. There is reason to believe that by proper selection of germplasm, four-wing saltbush plants can be selected for many arid zone sites. This genetic plasticity is important to future breeding and selection programs. Although the evolutionary patterns in *Atriplex* are still not clear, chromosomal races of diploids, tetraploids, hexaploids, and dodecaploids are well recognized (Stutz and Sanderson 1979).

THE NEW BIOTECHNOLOGY

Recent advances in plant tissue culture and molecular biology suggest that the domestication of *Atriplex* may be possible in a rather short period. Mass clonal propagation from meristem cultures is easily achieved, and somatic embryogenesis has recently been made possible. It would appear that somaclonal variation in cell suspension cultures is great enough to lead to an enormous germplasm diversity, and stresser systems may be imposed on those cells to select against specific environmental stresses. Once cells have been selected, somatic embryos can be regenerated, and whole plants can be reproduced and tested under field conditions. These experiments are now in progress, and we have every reason to believe that they will be successful.

The importance of the correct germplasm for revegetation of degraded landscapes and disturbed sites cannot be overemphasized. Since the areas in question are so large, and since the capital investment of revegetation is so great, one cannot afford to plant the wrong material. Selection of proper genetic strains for specific sites would appear to be on the horizon, and it may indeed be possible to select for high-yielding, palatable forage for hostile arid environments where no vegetation now exists. In addition to these advances, the potential for genetic engineering of range plants could bring new productivity to regions never before considered to have biological potential.

THE MANAGEMENT PROBLEM

Once desirable plants have been established, they must be protected from overgrazing or the experiment will be a failure. Effective systems of grazing management will vary throughout the world, but management is the key to success. It makes no sense to make a major capital investment in plant establishment, only to see the system fail because of overgrazing. There is, indeed, reason to be optimistic that biological productivity can be returned to degraded landscapes. Our best scientific knowledge must be incorporated to see that arid lands assume their share of providing for their own inhabitants. With sound ecology and management, the desert can bloom with a minimum of energy input.

LITERATURE CITED

Chapman, V. J. 1975. The salinity problem in general, its importance, and distribution with special reference to natural halophytes, p. 7–24. *In:* A. Poljakoff-Mayber and J. Gale (eds), *Plants in saline environments.* Springer-Verlag, New York.

Chew, R. M. and A. E. Chew. 1965. The primary productivity of a desert shrub (*Larrea tridentata*) community. *Ecol. Monographs* **35** (4): 355–375.

Goodin, J. R. and R. H. Newton. 1981. Unconventional plants for biomass feedstocks in semi-arid West Texas. Rept. SPI, 80-B-6, Texas Energy and Natural Resources Advisory Council, Austin, 45 pp.

Kelly, D. B., J. R. Goodin and D. R. Miller. 1981. The biology of *Atriplex*, p. 79–107. *In:* D. N. Sen and K. S. Rajpurohit (eds), *Contributions to the ecology of halophytes.* W. Junk, The Hague.

Laetsch, W. M. 1974. The C_4 syndrome: A structural analysis. *Ann. Rev. Pl. Physiol.* **25:** 27–52.

Lavrenko, E. M., N. V. Andruv and V. L. Leontief. 1955. Profile of the U.S.S.R. from the tundra to the deserts (translated title). *Botanicheski Zhurnaal U.S.S.R.* **40:** 415–419.

McKell, C. M. 1976. Shrubs – a neglected resource of arid lands. *Science* **187:** 803–809.

Mozafar, A. and J. R. Goodin. 1970. Vesiculated hairs: A mechanism for salt tolerance in *Atriplex halimus* L. *Plant Physiol.* **45:** 62–65.

Odum, E. P. and H. T. Odum. 1959. *Fundamentals of Ecology.* Saunders, Philadelphia.

Stutz, H. C. and S. C. Sanderson. 1979. The role of polyploidy in the evolution of *Atriplex canescens*, p. 615–621. *In:* J. R. Goodin and D. K. Northington (eds), *Arid land plant resources,* Intern. Center for Arid and Semi-Arid Land Studies, Texas Tech Univ., Lubbock, Tx.

Waisel, Y. 1972. *Biology of halophytes.* Academic Press, New York.

Forage production: a step towards better management of arid and semi-arid rangelands in countries of North Africa and West Asia

AHMED E. OSMAN

International Center for Agricultural Research in the Dry Areas (ICARDA), P.O. Box 5466, Aleppo, Syria

ABSTRACT

In many countries of West Asia and North Africa, rangeland served over the years as the main source of animal feed supply. At present, however, these rangelands are suffering great deterioration which has made them incapable of producing sufficient feed for the ever increasing livestock population. Feed shortage is especially acute during late summer and early winter. Under these constraints livestock production in the region is not increasing sufficiently to meet the demand for animal protein.

It is widely believed that providing alternative feed supplies from areas outside the range is highly important not only as a practical way of improving livestock production but also as a prerequisite to alleviating grazing pressure on the national rangelands. Great potential exists in the region for pasture and forage production, especially under rainfed agriculture, where over 40 million ha are estimated to be left fallow every year. The potential of using the fallows for pasture and forage and hence contributing to animal feed and to the stabilization of agriculture in the region is discussed together with its impact on rangelands.

Research by the Pasture Forage and Livestock Program at ICARDA is focussing on several legumes, and their management and utilization, with the objective of developing an effective rainfed crop-livestock system. Suitable pasture legumes have been identified such as *Medicago rigidula*, which is well adapted to the soils and climate of north Syria. The program has also extensively developed hay-producing systems using species of *Pisum* and *Vicia*. In countries of the region where such a system cannot be applied due to limited rainfall, irrigated forages should be encouraged wherever possible.

INTRODUCTION

The degradation of natural grazing land in countries of West Asia and North Africa is well documented (Pearse 1971, FAO 1972, Carter 1978, Huss 1978). Over the years these rangelands used to provide the bulk of animal feed requirement. The percentage of forage for livestock contributed by rangelands has been estimated at 90% in Iraq and Sudan, 86 to 88% in Jordan, Lebanon and Iran, 70 to 80% in Pakistan, and 60% in Syria (FAO 1972). However, several changes have taken place with time which have altered the traditional but efficient way of utilizing these rangelands. Some of the important changes are related to the increase in livestock number which has occurred all over the region. For example, increases amounting to 30–50% in Pakistan and Turkey, 176% in Sudan, and 225% in Egypt have been recorded. Such an increase is attributed on the one hand to successful livestock disease control measures and on the other to traditional behaviour of nomadic and tanshumant livestock owners who, in order to cope with the erratic climatic and environmental conditions, keep high livestock numbers in good rainfall years to compensate for high death losses in drought years. Under this situation, and in the absence of adequate feed supplies from other sources, rangelands have become overstocked beyond their carrying capacity, which in turn has accelerated their deterioration.

Ironically, the increase in livestock numbers has not resulted in sufficient increases of animal products which are still in great demand (Table 1). It is believed that at the present growth rate in livestock production (1–2%), most of the countries of the region will have to rely on importing these products, unless drastic measures are taken to improve livestock production. It is also widely believed that the most limiting factor for increasing livestock production is feed supply. Obviously rangelands cannot supply the

Table 1.* The gap between the consumption targets and production targets for animal products in individual countries in 1975 and 1985 (1,000 metric ton).

	1975			*1985*		
	Meat and offals	*Eggs*	*Milk*	*Meat and offals*	*Eggs*	*Milk*
Afghanistan	29	—	264	44	—	416
Iran	30	—	264	28	1	558
Iraq	27	—	144	55	(+1)	335
Jordan	13	1	70	18	—	120
Kuwait	23	2	102	35	3	157
Lebanon	44	(+4)	112	57	(+10)	172
Saudi Arabia	35	2	158	42	2	274
Sudan	(+7)	—	125	(+28)	—	59
Syria	4	1	118	—	1	244
U.A.R.	75	12	640	157	24	1285
Yemen	6	—	28	6	—	48
Regional total	279	14	2025	414	20	3668

*Source: FAO (1972).

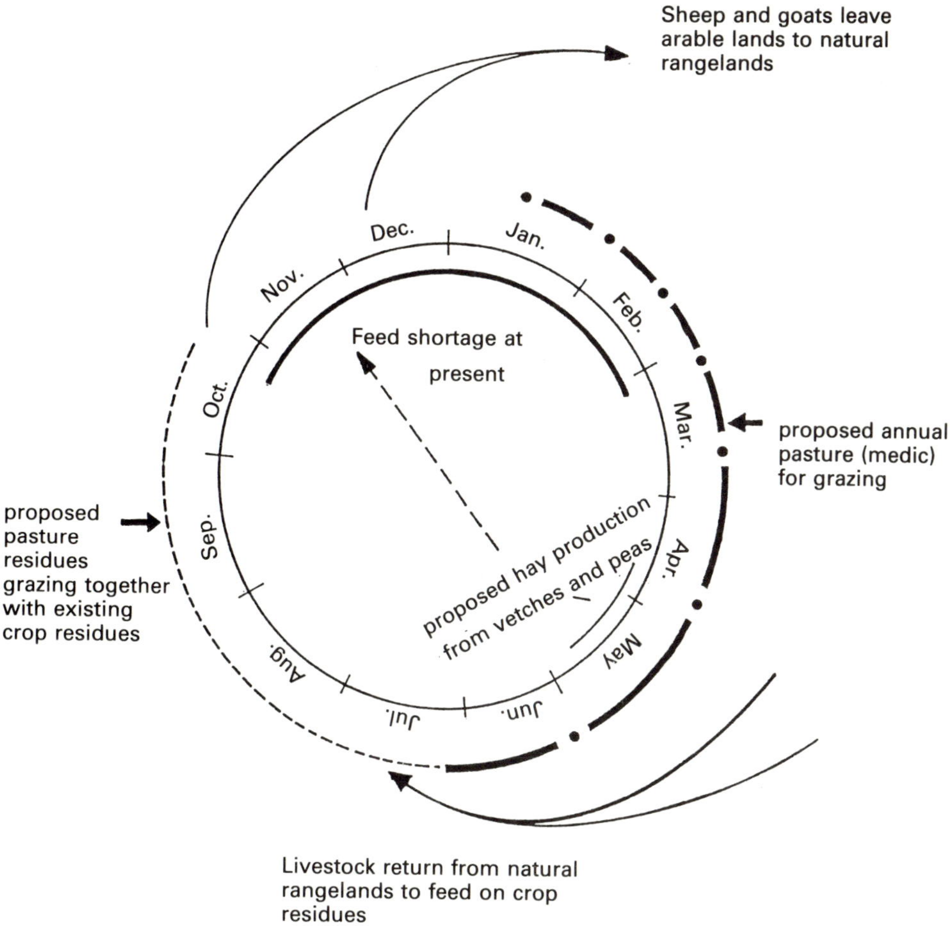

Fig. 1. Feed shortage, livestock movements and prospects of improving feed supply on arable lands.

required feed and so emphasis will have to be placed on feed sources from outside the range.

In many countries of the region, feed shortage is most acute during late summer and early winter. In Syria, for example, sheep and goats leave the cultivated areas in late summer for the rangelands (Legislative Decree No. 65 of 1966) and stay there until crop harvest when they are allowed to return to the cropping areas to feed on crop residues (Fig. 1). One remedy to the feed shortage, which would also help alleviate grazing pressures on rangelands (especially during the unfavourable times referred to in Fig. 1) is to encourage intensive forage and pasture production on available arable lands.

Forage crops on arable cropping land

Great potential for forage production exists in the region under both rainfed and irrigated farming systems. Under rainfed conditions, fallows may account for 50% of the arable land and the remainder is sown to grain (ICARDA 1982a). Over 40 million hectares are estimated to be left fallow every year, most of which exists in the medium and low rainfall areas (200–400 mm). Apart from straw, which is often conserved and used for supplement during winter, the present system of farming provides little feed for animal (Huss 1978). The main advantages claimed for the fallow in cereal/fallow rotations are weed control, improved soil fertility, and moisture conservation for the next cereal season. However, recent studies by ICARDA in Syria have indicated that when there is less than 300 mm of rain, a "farmer managed" fallow does not conserve moisture (ICARDA 1982b). Weed control, on the other hand, can be adequately achieved by the use of chemical herbicides while soil fertility (soil nitrogen) can effectively be improved by growing suitable legumes.

At ICARDA the Pasture Forage and Livestock Program is focussing on several legumes, with the objective of developing an effective rainfed crop-livestock system. Some of these legumes are intended for annual rainfed pasture while others are for hay production.

Annual rainfed pastures

The annual rainfed pastures are defined here as self regenerating communities to be used primarily for grazing. A legume suitable for an annual pasture in a cereal/pasture rotation under rainfed conditions should have the following characteristics: (i) adaptability to rainfall and temperature, (ii) production of large quantities of seed, (iii) high proportion of the seed produced should survive the cereal year, and (iv) self-generation during the pasture year.

Several species of medic (genus *Medicago*) meet the above requirements. The principle underlying this system is that the medic seeds have a dormancy period of one year or more. Only a relatively low percentage of the seeds produced in the first year germinate during the second year, when winter cereals are grown. During the third year, after the winter cereal crop has been harvested, the medic seeds (which are in the ground) will germinate to produce the medic crop. This system will therefore continue without further resowing of medics for a number of years.

A system based on the introduction of species of medic (*Medicago* spp.) and clover (*Trifolium* spp.) in rotation with rainfed cereals, accompanied by heavy applications of phosphorus is widely used in Australia. It is estimated that this system now contributes feed supply to about 180 million sheep equivalent each year, in addition to its benefit to soil fertility estimated to be 70–100 kg N/ha annually (Carter 1975).

The replacement of fallow by legume pastures has been attempted with variable success in some countries of the region (Tunisia, Algeria, Morocco, Lybia, Cyprus and Iraq). Such attempts have been based upon the introduction of medic varieties from Australia. However, it is now realized that these varieties are suitable only in areas having mild winters with a fairly long rainy season, i.e. the coastal areas. For the interior, where the winters become increasingly cold, better adapted medic cultivars are

Table 2. Performance of several genotypes of *Medicago rigidula* compared with *Medicago truncatula* var. *Jemalong* at four sites in Syria in 1982–1983.

			Dry matter (kg/ha)			
	Rainfall		*Qamishly* *314.8 mm*	*Hama* *372.4 mm*	*Salamieh* *292.1 mm*	*Izraa* *366.5 mm*
	Av. max. temp.		*18.5°C*	*19.1°C*	*18.7°C*	*19.4°C*
	Av. min. temp.		*2.2°C*	*−0.5°C*	*−1.7°C*	*−1.9°C*
Species	ACC. No.	Origin				
Medicago rigidula	835	Syria	4671	6646	1686	2780
Medicago rigidula	1783	Turkey	445	6531	1705	2058
Medicago rigidula	1075	Turkey	4343	6621	1071	1586
Medicago rigidula	281	Turkey	3957	5242	1533	1718
Medicago rigidula	811	Syria	3268	6282	1717	4184
Medicago truncatula var. *Jemalong*		Australia	318	4465	780	2728
LSD (5%)			1547	1124	984	1556
			Seed production (kg/ha)			
Medicago rigidula	835		463	821	210	724
Medciago rigidula	1783		415	674	277	495
Medicago rigidula	1075		591	940	277	878
Medicago rigidula	281		412	684	383	489
Medicago rigidula	811		495	1083.	330	775
Medicago truncatula var. *Jemalong*			68	501	133	681
LSD (5%)			188	358	140	326

needed. Research by ICARDA in north Syria, has identified a local species of medic (*Medicago rigidula*) which has great ability to survive cold winters, form nodules in conditions in which Australian medics fail to nodulate, and produces large quantities of herbage and seed (Table 2). Similarly in Iraq, local ecotypes of *Merdicago polymorpha* were found to perform considerably better than the Australian cultivars. This suggests that more emphasis has to be placed on plant selection from within the region which is rich in pasture legumes.

The identification of the suitable type of legume is only the first step. Other important factors required for the successful adoption of the system include shallow tillage practices, proper phosphate fertilization and proper grazing management. Apart from these technical problems, there are other constraints which have to be overcome. These include: (i) lack of adequate and readily available supply of seeds of suitable cultivars, (ii) inadequate research effort by national programs to initiate and sustain an integrated crop-livestock farming system, (iii) traditional segregated animal and crop husbandries strengthened in many cases by separate ownership of arable land and livestock, and (iv) lack of an effective body of extension service. Efforts are therefore needed to multiply seeds of locally adapted cultivars and to strengthen the research and extension services as well as to provide proper incentives for farmers to establish the annual pastures on their land, especially those who do not own livestock.

Hay production

Improvement of animal feed in the region can also come from forages which are grown and harvested for hay. This is especially needed during summer and early winter when there is a great deficit in feed supplies. In some countries (e.g. Morocco, Syria, Egypt, Tunisia and Jordan) the straw of cereal crops is used in combination with concentrates for animal feed during this period. These concentrates, however, are expensive and in short supply throughout the region. Local hay offers a good source of energy to be included in the rations of ruminant animals. Reid (1982) indicated that hay as a source of energy is much cheaper than concentrates (cereal grain and oil meals). Although hay making is not a common practice in the region, the technique is simple and farmers can easily learn it. Hay does not deteriorate rapidly in storage and, depending on the group of animals, it can supply their requirements for energy, protein, vitamins and minerals. The object in hay making is to reduce the moisture content of the green forage crops sufficiently to permit safe storage without spoilage or serious loss of nutrients. In the green stage, when forages are cut for hay, legumes contain about 70 to 75% moisture while most grasses contain 60 to 65% moisture. These levels should be reduced to 18–22% to allow proper storage.

Several forage crops can be used for hay production, and many of them are well adapted to the region. Examples are species of the genera *Vicia*, *Pisum* and *Lathyrus*. In Syria, common vetch (*Vicia sativa*), narbon vetch (*V. narbonensis*), woolly pod vetch (*V. dasycarpa*) and forage pea (*Pisum sativum*) are well adapted to the cold winters and in spring they grow vigorously, producing herbage yields of 5 to 6 t/ha in areas receiving 300 to 400 mm of rainfall. Chickling vetch (*Lathyrus sativus*) can be grown in drier areas (200 to 250 mm). In Iraq, narbon vetch and woolly pod vetch were identified as the most productive of all vetches tested (Radwan and Al-Fakhry 1975). Similar information from Lebanon, Jordan, Tunisia and Morocco confirm that vetches are well adapted forage legumes under rainfed conditions in the region.

Forage legumes can be grown in mixture with cereals such as oats, barley or triticale. These cereals, when included, have often resulted in increased yields and made easier mechanical harvest of the crop. Although productivity of forages depends on climatic and soil factors, hay yields of 5 to 10 t/ha can be expected in areas receiving between 300 and 400 mm of rainfall. The dry sunny weather which prevails following the rainy season is also in favour of hay-making operations.

As with annual pasture, serious efforts have to be made for the successful adoption of the hay-making system. Major constraints still exist before the system becomes attractive and profitable for farmers. Providing incentives in the form of seeds, fertilizers and machinery has proved encouraging to farmers in the Gezira area (N. E. Syria). In Cyprus, a hay subsidy that amounts to 8 Cypriot pounds ($18) per ton of hay up to a maximum of 14 ton was introduced in 1977. As a result, the area of forage increased from 1488 ha in 1977 to 4324 ha in 1978 with a consequent reduction in imported barley and concentrates. Strengthening extension services to disseminate new techniques to farmers is necessary throughout the region.

Forage under irrigation

In countries with adequate water, there is a good chance for irrigated forage production. Forage under irrigation is produced on a large scale only in Egypt, where over one million hectares contribute about 70% of total feed available for livestock. In most countries of the region, less attention is paid to irrigated forages because of limited irrigation water or because of lack of interest in forages compared with cash or food crops. In Iraq the estimated area of irrigated forages is 100,000 ha, contributing 12.4% of total available animal feed; in Saudi Arabia irrigated forages provide about 11%, while in Yemen, Oman, Bahrain and Qatar, fodder from irrigated forages constitute about 4% of total feed resources.

Nevertheless irrigation projects are being developed in many countries of the region where forages, especially legumes, could be included as part of the crop rotation not only to provide animal feed but also to improve soil structure and fertility. In Saudi Arabia, a recent resource survey showed a great potential for increasing agricultural production in existing arable areas through improved water distribution and drainage systems. The Government plan is to increase the area under irrigation by 50,000 ha over the present area of 121,000 ha (Fisher 1979). Accordingly, four important projects were undertaken: the Al-Hasa irrigation scheme, the Faisal Model settlement scheme, Wadi Jizan Dam Project and Abha Dam Project, resulting in additional irrigated area of 30,000 ha.

At present, alfalfa is the most widely used forage crop under irrigation in Saudi Arabia and in the Gulf countries. In Qatar, the area under alfalfa was estimated at 20% of the total cropped area of 1,200 ha (FAO 1972). Similarly, in Abu Dhabi the crop is considered one of the most adapted forage crops under irrigation. In Syria, irrigated forages were estimated in 1979 (Ministry of Agriculture) to cover 5,000 ha consisting mainly of alfalfa, but an additional 50,000 ha are considered possible from the Euphrates Project which is expected to allow for irrigation of 640,000 ha by 1990.

Impact of forage production in the region

It appears from the previous discussion that the highest potential for forage production exists in dryland farming. Carter (1978) in his survey of 9 countries in the region (Near East, North Africa) reported that 30 million hectares of land are left fallow every year under dryland farming. Assuming that 23 million hectares of this area are grown with suitable pasture and forage crop legumes, he estimated an increase in feed supply of about 68 million metric tons of dry matter (DM) (Table 3), which can support over 86 million ewe equivalent for a whole year. He also estimated an annual increment of 1.4 million ton of N (65% greater than total N

Table 3.* Existing fallow area, use of nitrogen fertilizer and estimated potential nitrogen increment and livestock feed supply from using pasture and forage legumes under rainfed conditions.

	Existing			*Potential*		
Country	*Total fallow area (10^6 ha)*	*Fertilizer N (ton)*	*Legume area (10^6 ha)*	*N-increment (ton)*	*Feed supply (10^6 metric ton DM)*	*Ewe† equiv. (10^6 ha)*
Afghanistan	4.0	27,668	2.4	144,000	6.0	7.5
Algeria	3.0	62,800	2.5	150,000	10.0	12.5
Iran	7.0	194,480	5.0	300,000	13.5	16.9
Iraq	2.0	25,000	1.5	90,000	6.0	7.5
Jordan	0.2	1,900	0.15	9,000	0.45	0.6
Libya	0.5	15,000	0.4	24,000	1.2	1.5
Syria	2.5	42,000	2.0	120,000	7.0	8.8
Tunisia	1.5	18,000	1.2	72,000	4.8	6.0
Turkey	9.0	452,620	8.0	480,000	20.0	25.0
Total	29.7	839,468	23.15	1,389,000	68.95	86.3

* Source: Carter (1978).
† Calculated assuming 800 kg DM/ewe/year.

fertilizer used in 1975) would be added to the soil from biological nitrogen fixation, and which would benefit cereal crops in the rotation. Moreover, he estimated that over 20 million ton of forage DM could be produced under irrigation, supporting an additional 26 million ewe equivalent/year. He concluded that animal feed can be greatly increased in the region without necessarily reducing the production of cereal and food legume crops.

Forage production, as outlined in the above discussion, is expected to have a positive effect on rangelands by: (a) making it possible to reduce overgrazing problems, and (b) allowing for adoption of proper grazing systems. Overgrazing, which is the result of the livestock population being greater than the carrying capacity, is considered the number one problem of rangelands in the region. With these forages, it should be possible to take large numbers of animals off the range to be raised entirely on arable land. Huss (1978) suggested that the overgrazing problem could be controlled if a land use system is developed whereby rangelands become the breeding grounds, and croplands the fattening grounds. He added that this would allow greater numbers of animals to be taken off the range leaving only the reproducing females. Of course, some people would argue that breeding animals require a level of nutrition (before and after lambing/calfing) that cannot be adequately supplied by rangelands in their present condition. However, the fact that large numbers of animals are to be removed from the range should leave behind adequate feed for the others, and if nutritional problems still prevail they can be solved by adequate supplementation.

At present, livestock move to the range at the beginning of the rainy season (Fig. 1) causing great damage to the newly emerging or sprouting vegetation through repeated defoliation and trampling. This will not be the case if feed in the form of conserved hay or medic pasture is made available on arable land, to support animals well through the winter and even beyond. Recent studies by ICARDA indicate that adapted medic species such as *Medicago rigidula* and *M. noeana* can produce between 40 and 50 kg DM/ha/day during winter which can easily support 10–12 sheep/ha. Spring productivity of the same pasture was between 100 and 160 kg DM/ha/day giving enough feed for 12 sheep/ha and at the same time allowing for high seed production. The excess spring production could provide for an additional 100 days of grazing in summer, thus carrying 12 sheep/ha for most of the year.

The point to be stressed here is that in the presence of adequate feed supply on arable land it should be possible to adopt a deferred grazing system whereby the range is allowed to set adequate seed and still be used for limited grazing.

LITERATURE CITED

Carter, E. D. 1975. The potential role of integrated cereal-livestock systems from Southern Australia in increasing food production in the Near East and North African Region. FAO/UNDP Regional Meeting Lahore-Peshawar-Karachi, Pakistan, 1975.

Carter, E. D. 1978. Legumes in farming systems of the Near East and North African region. Report to ICARDA, publ. by the University of Adelaide, Australia.

FAO. 1972. Near East regional study on animal husbandry, production and health. Fodder production and range management in the Near East and FAO's policies and plans for promoting the animal industry. FAO, Cairo.

Fisher, W. B. 1979. *The Middle East and North Africa 1979–1980. Europa Publ. Ltd., London.*

Huss, D. L. 1978. Importance of range development in dryland systems of farming for integration of crops and livestock husbandries in the Near East. Technology for increasing food production: Proc. 2nd FAO/SIDA Seminar on Field Food Crops in Africa and Near East, Lahore, Pakistan, 18 September to 5 October 1977.

ICARDA (International Center for Agricultural Research in the Dry Areas). 1982a. Promising medic cultivars for crop rotation, ICARDA Research Highlights: 54–56, Aleppo, Syria.

ICARDA (International Center for Agricultural Research in the Dry Areas). 1982b. Farming systems research, ICARDA Annual Report: 3–53, Aleppo, Syria.

Pearse, C. K. 1971. Grazing in the Middle East: Past, present and future. *J. Range Manage.* **24:** 13–16.

Radwan, M. S. and A. K. Al-Fakhry. 1975. The value of vetches for forage production in northern Iraq. *Mesopotamia J. Agri.* **10:** 21–31.

Reid, J. T. 1982. Quality hay, p. 532–547. *In:* M. C. Health, D. S. Metcalfe and R. F. Barnes (eds), *Forage, the science of grassland agriculture.* Iowa State Univ. Press.

Forage production and utilization of northern Mexico desert rangelands

MARTIN H. GONZALEZ

Instituto Nacional de Investigaciones Forestales y Agropecuarias, CIFAP-COAHUILA, Apartado 150, Saltillo, Coah., Mexico

ABSTRACT

Being geographically located at the same latitude, the deserts of northern Mexico and of Kuwait present some ecological similarities. The physiography is almost identical in many areas; annual rainfall varies from below 50 mm (Altar) to 95 mm (Vizcaino) to 220 mm (Samalayuca); maximum temperatures reach 42°C in the hot summer months. Vegetation is characteristic of these xeric environments, with alternate dominance of trees, shrubs, cacti, annual and perennial grasses, herbs and forbs growing in habitats ranging from sand dunes to rocky hills and mountains.

The Vizcaino, Altar, Samalayuca and Chihuahua deserts in Mexico are ecological regions within two of the greatest north American deserts: the Sonoran and the Chihuahuan-Coahuilan deserts, which occupy over 30 million hectares in eight western states.

Significant advances have been made in the inventory and detailed evaluation of Mexico's rangeland resources. However, in recent years, special emphasis has been placed on agronomic improvement of those resources in trying to make more efficient use of the limited and erratic rainfall, controlling undesirable plant species, and determining adequate utilization practices which include different intensities and systems of grazing with cattle, sheep and goats. These actions are based on the fact that, from the total average rainfall recorded annually, only 46% is finally available for desirable forage species; the other 54% is lost as run-off and evaporation, and utilized by undesirable plants; another small percentage falls outside the growing season.

In the extreme arid areas, e.g. the Altar Desert in Sonora, range improvement is almost limited to implementing flexible, moderate stocking rates under seasonal grazing patterns.

Forage production and utilization of the main desert plant communities in Mexico, as well as some aspects of general range management in these areas, are discussed in this paper.

INTRODUCTION

Perhaps no other land category in the world offers such large and numerous challenges to agriculturalists and pastoralists than the deserts. The unfavourable climatic conditions and, specifically, the very limited water resources, both above the ground and in subterranean acquifers, make any kind of life difficult. A special breed of people, plants and animals is needed to make these limited resources produce.

Attempts to understand our desert resources better and to exchange ideas and experiences on how to increase their productivity, are the reasons for holding the First International Conference on Range Management in the Arabian Gulf.

Arid and semi-arid lands in Mexico occupy about 50% of the country; this is equivalent to about 1,000,000 km^2. Within this extensive surface area, desert rangelands in northern Mexico occupy about 60 million hectares in eight states. These areas have many ecological characteristics similar to the desert ranges of Kuwait and other countries of the Arabian Gulf. Both are located around Lat. 30° N and have low rainfall (30–270 mm). The habitat ranges from sand dunes to rocky hills and elevated mountains. The typical xeric vegetation is dominated by sparse trees, shrubs and cacti, with some annual and perennial grasses and forbs in sites of higher rainfall and better soils.

For the purpose of this paper, the discussion will be centered around four of the Mexican deserts located in the arid northern part of the country, with many similarities also to parts of the southwestern United States. These deserts are encompassed within the two great deserts in North America, the Sonoran and the Chihuahuan deserts (Fig. 1). These areas have been selected because of their relative similarity to many arid rangelands in the Arab countries.

THE NORTHERN MEXICO DESERT RANGELANDS

Two of the great deserts in the North American continent are the Sonoran Desert, spreading from Arizona to the whole peninsula of Baja California and most of the State of Sonora, in Mexico, and the Chihuahuan desert (Fig. 1), covering parts of New Mexico and Texas in the United States, and great portion of the states of Chihuahua, Coahuila, Durango, Nuevo Leon and San Luis Potosi in Mexico.

The two drier regions within the Sonoran Desert have been chosen for this paper: Vizcaino in central Baja California, and Altar, in northwest Sonora. Within the Chihuahua Desert, the Samalayuca and the *Larrea-Fluorensia* regions in Chihuahua are discussed.

A. The Sonoran Desert

With a surface area of 310,361 km^2 in two countries, the Sonoran Desert occupies 188,926 km^2 (61%) in Mexico. Although a considerable part of the

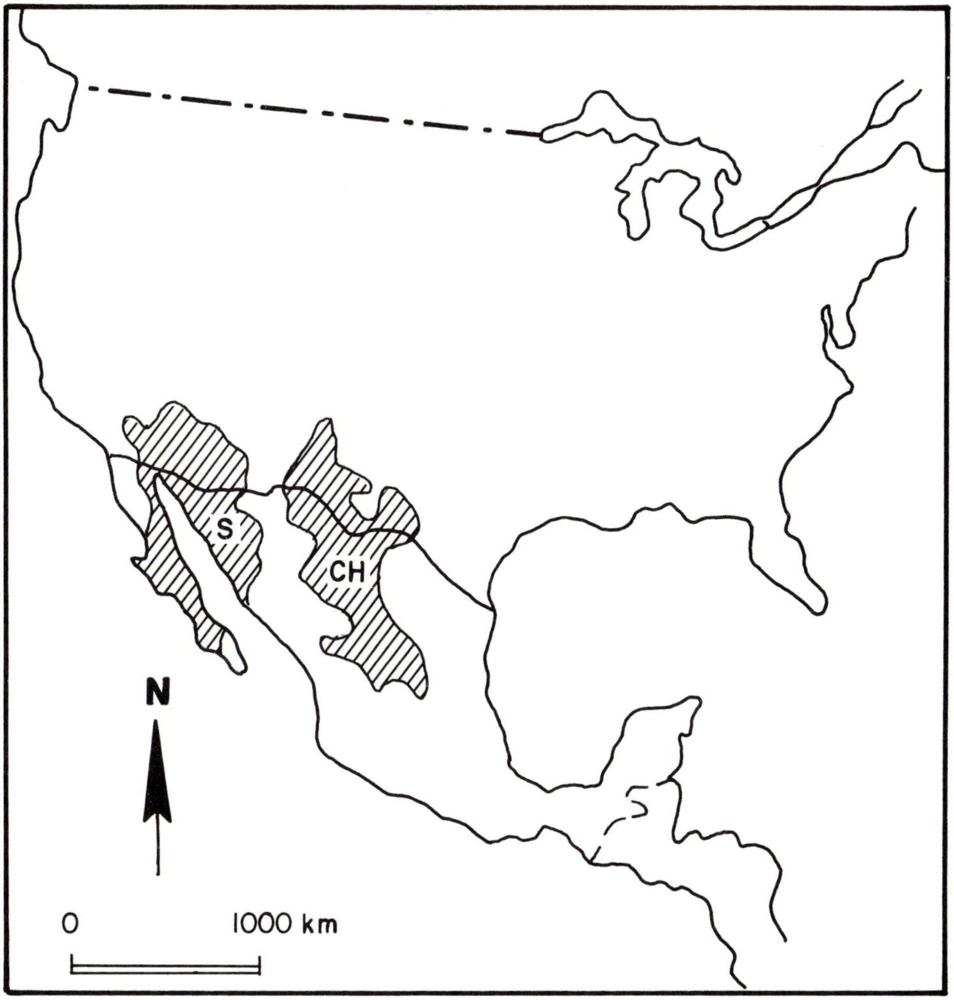

Fig. 1. Boundaries of the Sonoran (S) and the Chihuahuan (CH) Deserts in North America.

Sonoran Desert faces the Pacific Ocean, the Gulf of Cortes is nevertheless its dominant geographical feature, as well as the recipient of a very high percentage of its drainage (Shreve and Wiggins 1964).

Rainfall, amount and seasonal distribution, is the most important physical condition limiting the boundaries of the Sonoran Desert. In parts of Sonora and the peninsula of Baja California, the annual rainfall averages from 90 to 125 mm. However, some isolated areas in the Altar and Vizcaino Deserts never receive any precipitation at all. Other areas register between 10 and 50 mm (Martinez-Balboa 1981). According to Shreve and Wiggins (1964) ". . . of particular importance to plants in these areas are the number and duration of drought periods. Such periods, with a duration from 30 to 60 days, are of almost annual occurrence in the Sonoran Desert. The large non-succulent perennials are able to withstand them by drawing on the deep-seated moisture". Winter rains account for 40 to 50% of the annual total in southwest Sonora and the peninsula. The winter period extends from December to April, with rainfall of low intensity and longer duration than the summer short-duration storms of central Sonora. Most of the remaining precipitation falls in late summer, during September and October. Thus, there are two "peaks" of precipitation in this region.

Temperature is less variable throughout the desert than precipitation. Mean temperature averages around 22°C, with average maximum of 28°C, and average minimum of 17°C. However, periods of 90 days with a maximum of 40°C are not exceptional. Length of the growing (frost free) period ranges from 8 to 12 months in different parts of the desert. The duration of the potential growing season is interrupted by the drought periods, which sharply reduce all plant activity (Shreve and Wiggins 1964).

Soils are very low in organic matter due to the thin vegetative cover in general. Nitrogen and phosphorus contents are low, but calcium, magnesium and readily soluble salts are high. A calcareous hardpan (caliche) is frequently found 15 to 70 cm below the surface (COTECOCA 1981).

The vegetation of the Sonoran Desert has been extensively studied by Shreve and Wiggins (1964). Trees, shrubs and succulent plants are the main components of the vegetation, with change of dominance according to topography, soil and rainfall. Trees from the genus *Acacia*, *Bursera*, *Lysiloma* and *Prosopis* may reach 5 to 6 m, but more than half of the desert is dominated by shrubs of less than 1.5 m. However, *Larrea*, *Atriplex*, *Franseria*, *Lycium*, *Acacia*, *Caesalpinia* and numerous other genera may vary from 0.5 to 1.5 m in height. Ephemerals belong mainly to the genera *Daucus*, *Plantago*, *Pectis*, *Verbesina*, *Baileya* and *Tidestromia*. Succulents include *Opuntia*, *Pachycereus*, *Carnegiea*, *Lemaireocereus* and *Mammillaria*, among others. The few grasses found in the desert belong to the genera *Hilaria*, *Bouteloua*, *Cathestecum*, *Cenchrus*, *Tragus*, *Setaria*, *Aristida*, *Leptochloa*, *Scleropogon* and *Paspalum*.

B. The Chihuahuan Desert

The Chihuahuan Desert is separated from the Sonoran Desert by the West Sierra Madre, and covers an area of about 550,000 km^2 in the states of New Mexico and Texas (USA) and of Chihuahua, Coahuila, Durango, Zacatecas, Nuevo Leon and San Luis Potosi (Mexico). Elevation ranges from 1,100 m in the northern parts to over 2,000 m in the central plateau of Mexico (Rzedowski 1978).

Rainfall varies from 180 to 280 mm annually, being lower in the north and increasing southwards. About 80% of this rain falls during the late summer and early fall months. Snow storms are frequent in the northern states every year; freezing temperatures begin in late November and last until mid April, leaving a frost-free period (growing period) of 226 days as an average (Alvarez 1979). The characteristic plant species dominating the landscape of the Chihuahuan Desert are *Larrea tridentata* and *Flourensia cernua*.

Although most of this desert spreads over plains and low rolling hills, there are numerous calcareous rocky hills or small sierras throughout the area. These sierras support a crassicaulescent vegetation with *Yucca*, *Agave*, *Fouqueria*, *Dasylirion*, *Parthenium* and *Acacia* dominating a lower stratum of perennial grasses of the genera *Bouteloua*, *Tridens*, *Heteropogon* and *Aristida*, and some succulents like several species of *Opuntia* (prickly pear) (Hernández and Gonzalez 1959, Rzedowski 1978).

Other habitats of the Chihuahuan Desert include the drainage basins and closed watersheds in the lower sites. These sites are dominated by perennial grasses like *Hilaria mutica* or *Sporobolus airoides* in open stands or associated with *Prosopis juliflora* on slighty alkaline clayey soil. These *bajios* (low sites) are very productive under good management, with forage yields well above 650 kg/ha dry matter per year in years of normal rainfall (Gonzalez 1963, COTECOCA 1978).

Associated with the *Larrea-Flourensia* community, are other shrubs common in many sites; some of the most important are *Prosopis juliflora*, *Acacia constricta*, *Mimosa biuncifera*, *Parthenium incanum*, *Celtis pallida* and *Fouqueria splendens*. Among the grasses under the shrubs, the more common perennials are *Muhlenbergia porteri*, *Scleropogon brevifolius*, *Aristida* spp., and *Setaria macrostachya*, while annuals include *Aristida* and *Chloris* spp. (Hernandez and Gonazalez 1959, COTECOCA 1978, Valdez *et al.* 1975).

A very special vegetational subdivision of the Chihuahuan Desert, is the Samalayuca sand dune desert in the northern part of the state of Chihuahua. Vegetation in this zone is an interesting association of shrubs and perennial grasses, plus the flourishing ephemerals that provide an important component of the diet of cattle during the rainy summer season. Shrubs are mainly *Prosopis*, *Atriplex* and *Artemisia*, whereas the most important perennial grasses are *Sporobolus* and *Panicum* (Valdez *et al.* 1975, Rzedowski 1978, Marroquin *et al.* 1964).

FORAGE PRODUCTION AND UTILIZATION OF DESERT RANGELANDS IN NORTHERN MEXICO

The general ecological characteristics described for the Sonoran and the Chihuahuan Deserts indicate some of their similarities to the deserts of the countries of the Arabian Gulf. But if some similarities exist in ecological aspects, the utilization systems are different: nomadism in one country, and rural settlements in the other. But whatever the traditional pastoral systems in our countries may be, a deep understanding of the basic principles of how plants react to the environment, and to the pressure exerted upon them through defoliation by herbivores, is absolutely necessary if a sustained productivity is to be achieved. Fig. 2 indicates the distribution of the local deserts in northern Mexico.

A. The Vizcaino Desert

Located in the central part of the peninsula of Baja California, the Vizcaino region occupies a surface area of 35,000 km^2. Five vegetative

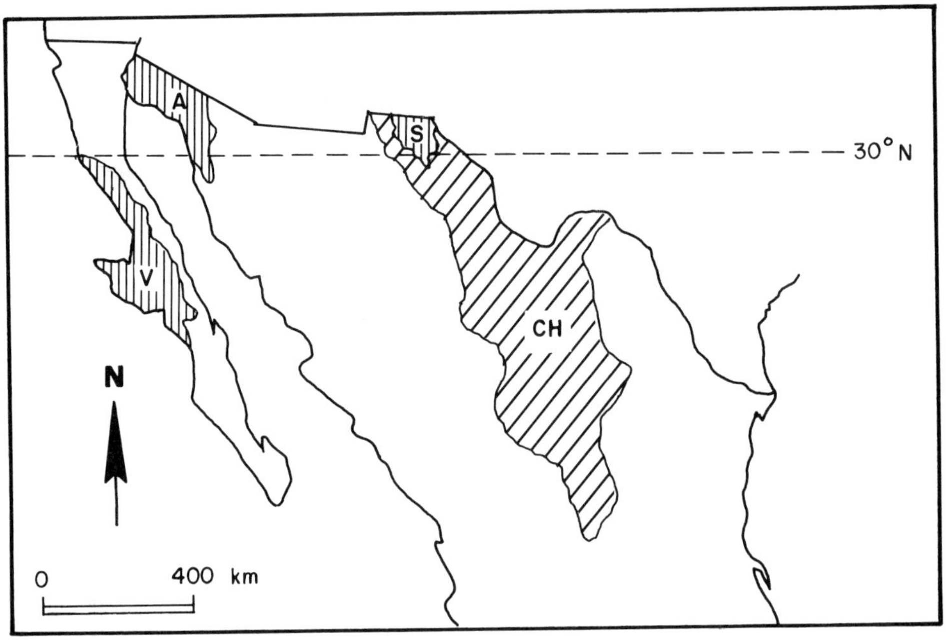

Fig. 2. Distribution of the Vizcaino (V), Altar A), Samalayuca (S), and *Larrea-Flourensia* (CH) Deserts in northern Mexico.

subdivisions of the Sonoran Desert spread over that area, but the sarcophyllous type is the most common in the gentle slopes and plains along the central and western portions of this desert (Shreve and Wiggins 1964).

Table 1 shows the main physical characteristics of Vizcaino, and the principal forage plants normally used by livestock (Martinez-Balboa 1981, Gonzalez 1983a). Annual rainfall ranges from 55 mm in Emiliano Zapata to 119 mm in Josefa Ortiz, to 203 mm in San Bruno, with an average in the five locations of 110 mm per year. There are two peaks of precipitation, one in December–January and the other in August–October.

Forage production and utilization

The main forage plants in Vizcaino cover different strata, but usually with a low basal density. Woody species mostly used by animals (cattle and goats) are *Cercidium floridum*, *C. microphyllum*, *Acacia californica*, *A. peninsularis*, *Fouqueria peninsularis*, *Franseria chenopodifolia*, *Lycium californicum*, *Prosopis juliflora* and *Atriplex polycarpa*. Management of these species includes, besides the direct browsing by animals, the pruning of branches not reached by cattle or goats, so that the green parts, mostly of *Cercidium*, *Leuchaena* and *Prosopis*, can be eaten at floor level (Gonzalez 1983b). A similar process is practiced with tall succulents like *Pachycereus pringleii* and *Lemaireocereus thurberi*. The heavy "arms" of these plants are cut and fed to animals. Apart from these arms, the flowers and the fruits of *Pachycereus*, falling naturally on the ground, constitute an important component of the animal diet.

Table 1. Main features and forage production of selected sites in the Vizcaino Desert, Baja California, Mexico.

Site	*Elevation (m)*	*Vegetation type*	*Annual rain (mm)*	*Main forage* spp.	*Forage prod. (kg/ha)*	*Carrying capacity (ha/AU)*	*Livestock*
San Bruno	0–15	Sarcocaulescent	203 (S-O) (D-J)	*Acacia* *Opuntia* *Fouqueria* *Pachycereus* *Cercidium* *Enneapogon* *Aristida* *Setaria*	59	62	Cattle Goats
San Ramon	60–700	Arbosuffrutescent	97.7 (A-O) (D-J)	*Cercidium* *Leuchaena* *Prosopis* *Opuntia* *Atriplex*	51	71	Cattle Goats
Cadeje	0–600	Sarcophyllous	76.4 (S-D) (D-J)	*Atriplex* *Cercidium* *Ephedra* *Acacia* *Bouteloua* *Enneapagon*	46	80	Cattle Goats
E. Zapata	240	Sarcophyllous	55 (A-S) (D-F)	*Atriplex* *Opuntia* *Pachycereus* *Plantago*	42	88	Cattle
Average			110		49	75.4	

An average production of 16 kg/tree of the tender tips of branches has been estimated for *Cercidium*, harvestable during the 10 to 12 weeks that green leaves are on the trees. With an average density of 12 trees/ha, the seasonal forage production for this plant is about 192 kg/ha (40% on dry matter basis) as reported by Gonzalez (1983b). For the tall, succulent *Pachycereus pringleii*, flowers falling to the ground during spring to early fall yield an average of 1.5 to 2.5 kg/plant, while for fallen fruits the yield varies from 2 to 5 kg/plant. The average population of 8 plants/ha thus produces about 16 kg of flowers and 28 kg of fruits per hectare that cattle eat readily.

The net edible forage produced by the harvested arms of *Pachycereus* would amount to 134 kg/ha/year. This is based on the assumption that only one arm of 50 kg average weight is cut per year, and the average density of mature plants is 8/ha. This would give an annual yield of 400 kg/ha. Converted to dry matter (40%) and deducting non-edible parts (16%), the net usable forage would be 134 kg/ha/year (Gonzalez 1983a).

Grasses are not frequent in Vizcaino, other than some annuals like *Aristida adscensionis*. However, in some years it is possible to find some perennials in areas not reached by livestock or fauna. These grasses are mainly *Setaria macrostachya*, *Enneapogon desvauxii*, *Digitaria californica* and *Bouteloua curtipendula*, particularly in the stony hills.

Research at COTECOCA, the National Institution for the Inventory of Rangeland Resources in Mexico, has led to the determination of the forage production in the Vizcaino Desert, including all types of plants and their availability through the year for cattle and goats (COTECOCA 1981). Table 1 shows also the forage production of five sites in Vizcaino, with an average of 49 kg/ha dry matter (varying from a low of 42 to a high of 59 kg/ha) in ranges in good condition and in years of normal rainfall. Estimated carrying capacity, therefore, varies from 62 to 88 ha/AU yearlong, averaging 75.4 ha/AU. Protein and phosphorous supplementation to animals is necessry during the dry period, from February to July (COTECOCA 1981, Gonzalez 1983b).

B. The Altar Desert

Altar (Fig. 2) is one of the deserts within the Lower Colorado Valley, the larger of the Sonoran Desert. It occupies the northwest corner of the state of Sonora, and is one of the most arid areas in all Mexico. Most of the area comprises plains and *bajadas* of low gradient, soils ranging from sand dunes to gravelly sites, at elevations from 0 to 180 m (Shreve and Wiggins 1964).

Rainfall in the Altar region varies from almost zero in the west to 165 mm in Sonoyta, in the eastern portion. Annual mean temperature is about 27.7°C, with frequent maximum of 40°C and minimum of 14°C. Table 2 gives the annual rainfall, key forage species, carrying capacity and forage production for four locations in the Altar Desert.

From the point of view of utilization by livestock, this area has a very distinctive characteristic. Because of its low winter-spring rainfall and its poor soils, pastoral activities are limited to 3 months during that season, when the ephemeral species grow. These ephemerals, which constitute the basis of the animal diet, are: *Plantago insularis*, *Cryptantha angustifolia*, *C. costata*, *C. holoptera*, *C. maritima* and *Erodium texanum*. Among the

Table 2. Main features and forage production of selected sites in the Altar Desert, Sonora, Mexico.

Site	*Elevation (m)*	*Vegetatin type*	*Annual rain (mm)*	*Main forage* spp.	*Forage prod. (kg/ha)*	*Carrying capacity** *(ha(AU)*	*Livestock*
San Luis		Microphyllous desert	55	*Plantago Cryptantha*	25	36	Steers (seasonal)
Riito	0–150	Microphyllous desert	40	*Ephedra Atriplex*			
Sonoyta		Microphyllous desert	165	*Krameria Hilaria*			
Sahuaro		Microphyllous desert	86	*Aristida Enneapogon*			

*Utilization during three months only; during the rest of the year no drinking water for livestock is available.

perennial shrubs found in scattered spots, are *Atriplex canescens*, *Beloporone californica*, *Krameria parvifolia*, *Ephedra trifurca* and perennial grasses like the tough *Hilaria rigida* and *Aristida californica*, *Heteropogon contortus* and the biennial *Enneapogon desvauxii*. Considering the production availability of these species, workers at COTECOCA (1980) determined a forage production of 25 kg/ha dry matter, with a grazing capacity of 36 ha/AU but only for 3 months, since no drinking water is available during the rest of the year. Feeder calves from central Sonora and other states, are taken first to the Altar Desert and then to the irrigated pastures in Mexicali, in the lower Colorado Valley.

C. The Samalayuca Desert

Samalayuca is a typical sand dune desert, located in the north central part of the state of Chihuahua, in an area of about 3,200 km^2. There are two types of sand dunes: large, more or less bare dunes, and low stabilized dunes covered with perennials and ephemerals.

Climate is dry, temperate, with hot summers and cold winters. Precipitation averages 193 mm, most of which falls during late summer and early fall. Annual mean temperature is 17°C, with extreme maximum up to 40°C in summer, and extreme minimum of −12°C in winter. Normally, there is a 10-month dry period, and a 210-day frost-free period (COTECOCA 1978).

Within the grass family, the most important species are the perennials: *Sporobolus flexuosus*, *S. airoides*, *S. contractus* and *Panicum harvardii*, and annuals from the genera *Aristida*, *Chloris*, *Bouteloua* and *Eragrostis*. Perennial shrubs include *Atriplex canescens* and *Prosopis juliflora* as the most palatable to livestock, and *Artemisia* spp., *Larrea tridentata* and *Ephedra trifurca*. Important components of the animals diet are ephemerals from the Plantaginaceae, Compositae and Onagraceae families, particularly the genera *Plantago*, *Aphanostaphus*, *Heterotheca* and *Oenothera*. Many other forbs and weeds like *Verbena*, *Baileya* and *Aster* dress the desert with a multicolour landscape during the summer.

Grazing by cattle in Samalayuca is a year-long operation, with *Atriplex* and the new re-growth of *Prosopis* in the spring, providing most of the protein in animal diet. However, the bulk of the daily consumption throughout the year depends on the perennial grasses. Two sites have been distinguished (COTECOCA 1978) on the basis of productivity, relief and soil conditions. These are the stable, low sand dunes and the tall somewhat dynamic sand dunes. Both shelter the same species of grasses, ephemerals and microphyllous shrubs but their density may be somewhat lower in the dynamic dunes. Table 3 shows that the two sites differ slightly in forage production and carrying capacity.

D. The Larrea-Flourensia *region*

This vegetational subdivision is the most extensive and the most typical representative of the Chihuahuan Desert, with its characteristic *Larrea*

Table 3. Main features and forage production of selected sites in the Samalayuca desert, Chihuahua, Mexico.

Site	*Elevation (m)*	*Vegetation type*	*Annual rain (mm)*	*Main forage* spp.	*Forage prod. (kg/ha)*	*Carrying capacity (ha/UA)*	*Livestock*
Chontes	1350	Shrubs on low dunes	205	*Sprobolus* *Panicum* *Atriplex* *Prosopis* *Aristida* Ephemerals	68	54	Cattle
Samalayuca	1290	Shrubs on high dunes	181	*Sporobolus* *Panicum* *Atriplex* *Chloris* *Chloris* *Krameria*	61	60	Cattle
Average			193		64.5	57	

tridentata-Flourensia cernua association. The distribution of this desert in northern Mexico has been mentioned in previous pages. The climate is very dry, temperate, with hot summers, although freezing temperatures are common from December to March. Mean annual temperature ranges beteen 15° and 22°C, and annual precipitation varies from under 200 mm in the northern part, where the almost pure *Larrea-Flourensia* association can be found, to over 350 mm in the southern part, where the two species are mixed with some thorny shrubs, forbs, cacti and grasses (COTECOCA 1978). Some of the most important associated species belong to the genera

Table 4. Main features and forage production of selected sites in the *Larrea-Fluorensia* region, Chihuahua, Mexico.

Site	*Elevation (m)*	*Vegetation type*	*Annual rain (mm)*	*Main forage plants*	*Forage prod. (kg/ha)*	*Carrying capacity (ha/AU)*	*Livestock*
Ahumada	1380	Microphyllous on plains	220	*Muhlenbergia* *Hilaria* *Scleropogon* *Sporobolus* *Prosopis* *Atriplex*	136	36	Cattle
Aldama	1460	Microphyllous on low hills	230	*Muhlenbergia* *Sporobolus* *Bouteloua* *Hilaria* *Atriplex* *Parthenium* *Prosopis* *Opuntia*	123	40	Cattle Goats Sheep
Average			225		130	38	

Acacia, *Mimosa*, *Ambrosia*, *Fouqueria*, *Zinnia*, *Agave* and *Opuntia*, most of which have some forage value (Rzedowski 1978).

Cattle and goats are the main domestic herbivores grazing or browsing the *Larrea-Flourensia* region on a year-long basis. Supplementation of protein and phosphorus is necessary during 5 to 7 dry months in order to obtain good fertility and market weights (Gonzalez 1973). Considering two of the most representative sites within this desert (Table 4), forage production averages from 123 to 136 kg/ha dry matter in years of normal rainfall in the Ahumada and Aldama sites. These production figures are equivalent to a carrying capacity of 36 to 40 ha/AU/year if the range is in good condition. A decrease in condition might reduce the capacity to 65–72 ha/AU/year (COTECOCA 1978).

RANGE IMPROVEMENT IN ARID ECOSYSTEMS

Considering the characteristics of the deserts in northern Mexico, and of other desert ecosystems in the world, it seems that the potential for improving their forage and animal production rests in three basic actions: (1) Try to make the best possible use of rainfall, reducing runoff and evaporation and increasing infiltration; (2) Eliminate competition from undesirable species without exposing the soil to erosion; and (3) Adopt better forage utilization systems of desert plants through animals or harvesting. A fourth action, revegetation of desert sites, is another alternative, limited at least for commercial large scale operations, to areas where rainfall is from 150 to 200 mm, with a more or less reliable distribution, and where edaphic conditions are suitable.

The priority of improvement practices lies in the manipulation of water uses. Fig. 3 illustrates the allocation of annual rainfall to the various habitat components in a *Larrea-Flourensia* community in the Chihuahuan Desert. Out of an average rainfall of 205 mm per year, losses by evaporation and runoff account for 100 mm (48% of the total). Rainfall in the winter months averages 25 mm (12%), so that the net balance available for all types of plants present is reduced to only 80 mm, or 39% of the recorded precipitation. For a range in good condition in the *Larrea-Flourensia* community, most of the vegetative cover is made up of undesirable shrubs, and thus the final amount of rain left for the desirable forage species is only 33 mm (17% of the total recorded precipitation). Since climate cannot be manipulated by man yet, the options available for improvement of rangelands have to be centered around the modification of the vegetative cover, so that the limited rainfall can be used mostly by desirable plant species. Successful practices such as brush control, reseeding or other modification of the vegetation must be integrated into an overall land management program based on in-depth understanding of all the ecological factors (Scifres and Merkle 1975).

Agronomic improvements of arid rangelands such as reseeding, offer a series of risks that must be considered seriously because of the high expenses ivolved in this practice. Nevertheless, encouraging results have been obtained in different locations, where reseeding was combined with

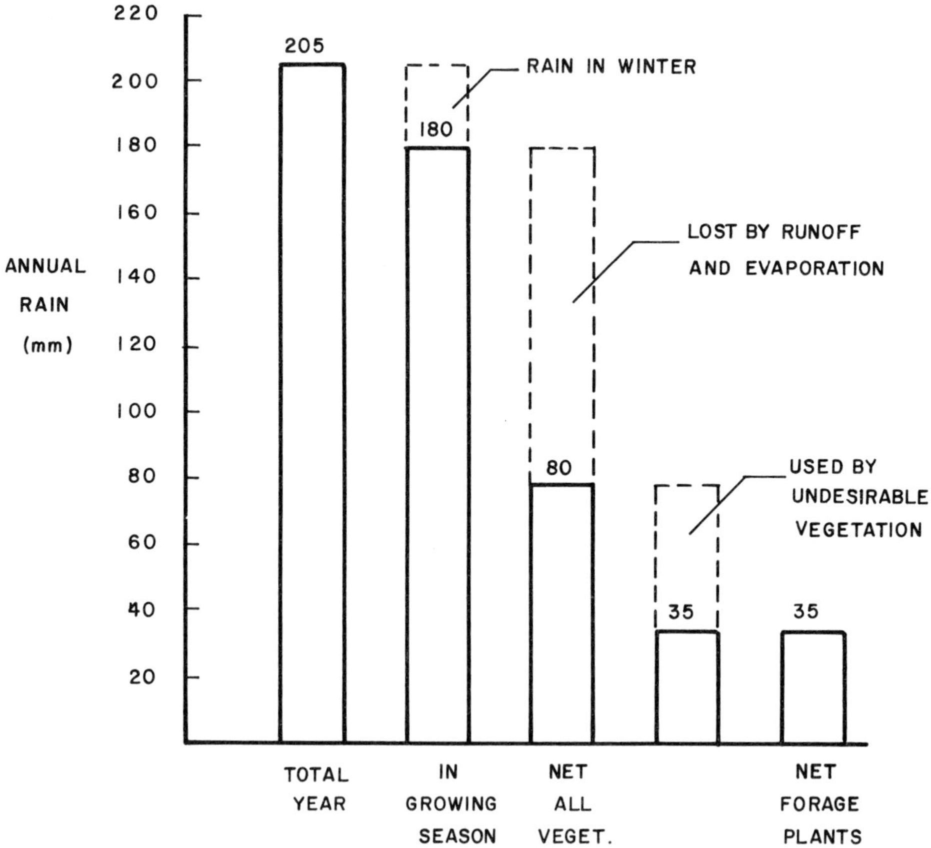

Fig. 3. Rainfall losses and use by plants in *Larrea-Flourensia* communities in northern Mexico.

water conservation practices (Gonzalez 1973, Javalera *et al.* 1976, Fierro *et al.* 1979, Sierra 1982, Tromble and Wood 1981, Wood and Buckhouse 1983). Reseeding of arid and semi-arid ecosystems has been extensively discussed by Herbel (1972, 1982) in New Mexico, and by other authors in different locations (Thomas and Young 1956, Anderson *et al.* 1957, Frost and Hamilton 1965, Branson *et al.* 1966, Slyback and Cable 1970). Common objectives and goals have been, among others, to eliminate or reduce the competition of noxious plants and substitute them by desirable species, reduce runoff and improve infiltration, and implement an adequate management plan after the improvement actions.

But if man has developed the methodology and equipment to face the challenge that the deserts offer, more in-depth evaluation and testing of new germplasm is necessary world-wide. This new germplasm must be able to withstand extreme drought conditions, high temperatures and evaporation. A more effective international exchange of germplasm material should be implemented, and a more active exchange of research results and experiences must be encouraged. Some plant material from the deserts of Vizcaino and Altar in Mexico may prove useful to the Arab countries, and some plants from the Arabian peninsula could be of great benefit to those areas with similar conditions in Mexico.

LITERATURE CITED

Alvarez, A. 1979. Boletín Meteorológico del estado de Chihuahua. Secretaría de Agricultura y Ganadería, y Gobierno del estado de Chihuahua, México.

Anderson, D., L. Hamilton, H. Reynolds and R. R. Humphrey. 1957. Reseeding desert grassland ranges in southern Arizona. *Arizona Agri. Exp. Sta.* Bull. 249.

Branson, F. A., R. F. Miller and I. S. McQueen. 1966. Contour furrowing, pitting and ripping on rangelands of the western United States. *J. Range Manage.* **19:** 182–190.

COTECOCA. 1978. Coeficientes de agostadero para el estado de Chihuahua. Secretaría de Agricultura y Recursos Hidráulicos, México, D. F., México.

COTECOCA. 1980. Coeficientes de agostadero para el estado de Sonora. Secretaría de Agricultura y Recursos Hidráulicos, México, D.F., México.

COTECOCA. 1981. Coeficientes de agostadero para el estado de Baja California Sur. Secretaría de Agricultura y Recursos Hidráulicos, México, D.F., México.

Fierro, L. C., J. Javalera, M. H. González and F. Ibarra. 1979. Comparación del establecimiento de cuatro mezclas de zacates nativos e introducidos en cuatro camas de siembra. *Rancho Exper. La Campana,* Boletín, Pasizales 10:6, INIP-SARH.

Frost, K. R. and L. Hamilton. 1965. Basin forming and reseeding of rangelands. *Trans. Amer. Soc. Agric. Eng.* **8:** 202–203.

González, M. H. 1963. Estudio integral de la ganadería del norte de México. Centro de Investigación del Desarrollo, Ed. COPARMEX, México, D. F., México.

González, M. H. 1973. El Plan Zacatecas de mejoramiento de pastizales. Primer Informe al Gobierno del Estado (Mimeo). Zacatecas, Zac. México.

González, M. H. 1983a. Proyecto de un sistema de producción agropecuaria para el ejido San Ramón, Comondu, B.C.S. SARH-OHIADR, La Paz, B.C.S., México (Mimeo).

González, M. H. 1983b. Proyecto de un sistema de producción agropecuaria para el ejido Emiliano Zapata, Mulege, Baja California Sur. SARH-OHIADR, La Paz, B.C.S., México (Mimeo).

Herbel, C. H. 1972. Using mechanical equipment to modify the seedling environment, p. 369–381. *In:* Wildland shrubs – their biology and utilization. USDA Forest Service Tech. Report INT-1.

Herbel, C. H. 1982. La resiembra de pastizales en ecosistemas áridos y semi-áridos. Memorias, XXV. Aniversario Rancho Exper. La Campana, I.N.I.P.-SARH, Chihuahua, México.

Hernández, H. E. and M. H. González. 1959. Los pastizales de Chihuahua. *Rancho Exper. La Campana.* Cir. 3, Chihuahau, México.

Javalera, J., V. Ortiz and F. Gómez. 1976. Efecto de la fertilización en el establecimiento de dos zacates nativos y cuatro introducidos. *Rancho Exper. La Campana,* Boletín Pastizales 8:3, Chihuahua, México.

Marroquín, J., G. Borja, R. Velázquez and J. A. de la Cruz. 1964. Estudio ecológico dasonómico de las zonas áridas del norte de México. Publicación Especial No. 2, Instituto Nacional de Investigaciones Forestales, México, D.F. México.

Martínez-Balboa, A. 1981. La ganadería de Baja California Sur. Vol. 1. Editorial J.B., La Paz, Baja California Sur, México.

Rzedowski, J. 1978. *Vegetación de México.* Editorial Limusa, S.A. México, D.F., México.

Scifres, C.J. and M. G. Merkle. 1975. Herbicides and good management improve range pastures. *Weeds Today* (Spring).

Shreve, F. and I. L. Wiggins. 1964. *Vegetation and flora of the Sonoran desert.* Stanford Univ. Press. 2 Vols.

Sierra, J. S. 1982. Establecimiento y productividad de cuatro mezclas de gramíneas en la región central de Chihuahua. *Rancho Exper. La Campana,* Boletín Pastizales 8:4, Chihuahua, México.

Slyback, R. D. and D. R. Cable. 1970. Larger pits and reseeding of semi-desert rangeland. *J. Range Manage.* **23:** 333–335.

Thomas, G. W. and V. A. Young. 1956. Range pitting and reseeding trials on the Texas Range Station near Barnhart, 1950–1955. *Texas Agri. Exp. Sta.* Prog. Rep. 1882.

Tromble, J. M. and M. K. Wood. 1981. Arid rangelands treatment effects on the control of erosion and sedimentation in rangelands improvement for New Mexico. New Mexico State Univ. Special Report 41, Las Cruces, New Mexico.

Valdez, J., A. A. Beetle and M. H. González. 1975. Gramíneas de Chihuahua. *Rancho Exper. La Campana,* Boletín Pastizales 6:3.

Wood, M. K. and J. C. Buckhouse. 1983. Technologies for capturing and retaining water on rangelands. *New Mexico Agri. Exp. Sta.* Special Report 52, Las Cruces, New Mexico.

Performance of three breeds of Arabian sheep under Kuwait's rangeland conditions

M. A. RAZZAQUE, F. K. TAHA AND A. R. SULEIMAN

Aridland Agriculture Department, Food Resources Division, Kuwait Institute for Scientific Research, P.O. Box 24885, Safat 13109, Kuwait

ABSTRACT

Rangelands form 90% of the total area of Kuwait. Several breeds of fat-tailed sheep are being raised here for meat, wool and milk production. Sheep represent about 80% of the total livestock population in Kuwait and provide about 5% of the total demand for red meat. Three major breeds of sheep (Naeemi, Nigdi and Erbi) are commonly raised in Kuwait and have become acclimatized to the weather and vegetation of the rangelands. Research data on the various performance parameters under different grazing systems are scanty and fragmented. Studies on the response of local sheep breeds in different grazing systems and seasons are essential for recommending suitable management practices in order to optimize animal production without degradation of rangeland resources. The present study was initiated to monitor the performance of three breeds of local sheep in different seasons and under two grazing systems.

Range sheep production studies were initiated at the Kuwait Institute for Scientific Research (KISR) in 1980–1981. Five paddocks (1 × 2 km each) were allocated for the grazing study within KISR's Sulaibiya site. One paddock (No. 1) was assigned for continuous grazing, three paddocks (Nos. 2–4) for rotational grazing and the fifth one for control. An area measuring 2.5 × 1 km was assigned as a holding pasture. In the first year, three breeds of sheep were selected consisting of 43 mature ewes from each of the Naeemi, Nigdi and Erbi breeds. These animals were grazed in the holding pasture with provision of limited amounts of supplement feed for 8 months (commencing July 1981). They were bred and introduced to continuous and rotational grazing systems allowing 5 ha/sheep for 122 days in the first year. The mean birth weight of lambs ranged from 4.03 to 4.81 kg depending on the breed. Lambing percentage varied appreciably due to breed differences. There was no significant differences between grazing systems with respect to response of ewes and young lambs.

During the second year, the sheep were provided with supplement feed from July to mid-December. An improvement in lambing percentage was observed especially in Nigdi and Erbi breeds, and the mean birth weight of lambs ranged from 4.26 to 4.38 kg. The duration of continuous and rotational grazing studies in the second year was 195 days and 3.45 ha were allowed per ewe. There was no significant effect on animal response due to type of grazing system. The breed differences in various peformance parameters of local sheep are discussed with reference to grazing during the different seasons of the year and grazing systems.

INTRODUCTION

In recent years, world meat consumption has increased more rapidly than production. The increase in meat consumption in oil-producing countries appears to be quite dramatic, as the estimated import of meat and live animals was about seven times more in the 1980s than in the 1970s. Kuwait and other Gulf countries depend mainly on imported meat or on live sheep for mutton. People in the Gulf traditionally prefer mutton and lamb; therefore, during the recent years the importation of live sheep has greatly increased since domestic production is negligible compared to demand.

In Kuwait, 80% of the total grazing animals are sheep and the rangeland forms about 90% of the country's total area. Several breeds of fat-tailed sheep have been raised by the people of the Arab arid regions including Kuwait for production of meat, milk and wool for many centuries. A traditional extensive system of sheep production has been practised in this region utilizing rangeland resources without any control in grazing resulting in degradation of rangeland resources. Studies conducted in Iraq on fat-tailed sheep (Demiruren 1972) showed that daily gain of Awassi sheep ranged from 103–244 g under feedlot operation and indicated that the lambing rate, carcass yield, milk and wool production could be greatly improved through changing husbandry practices. Research conducted on desert sheep in India (Mittal and Ghosh 1980) and in Iraq (Younis *et al.* 1977) showed encouraging results, particularly on the adaptability and tolerance of sheep under adverse conditions. Saleh *et al.* (1970) showed that improvement of weaning weight, daily gain and dressing percentage in Awassi sheep was possible by improved management practices such as controlled grazing and supplemental feeding. Indigenous sheep of Kuwait and neighbouring Gulf countries are known to have been adapted to the arid environment and harsh climate. However, research findings on the productive potential of these sheep with regards to their performance under the rangeland conditions of Kuwait during different seasons and grazing systems are scanty and fragmental.

Several studies carried out in the arid regions of the world, showed the profound effects of regional climate, especially heat stress, on the various performance parameters of the ewe, rams and lambs (Cartright and Thwaites 1976; Johnson and Hahn 1982). The potential value of Kuwait's rangeland and its vegetation for range sheep production has been reviewed by Taha and Krueger (1981). Various problems and advantages of animal production in the arid regions have been noted by Payne (1969). The main problems appeared to be associated with low rainfall, resulting in production of highly nutritious forage for a short period, followed by a long dry period during which forage growth ceases and vegetation dries up. The advantage in this situation is that the ruminant can graze almost through the year on standing hay in the rangeland. Payne (1969) indicated also, that plants in arid regions, when mature, become a poor source of crude protein, which results in the poor performance of grazing animals. In this situation, monitoring the plant quality and the animal response might be useful to make provision for supplemental feed when needed in order to maintain the productivity of grazing animals.

The present study was undertaken to investigate performance of three breeds of Arabian sheep under Kuwait's rangeland conditions during different seasons of the year.

MATERIALS AND METHODS

Field animal facilities

The experimental grazing area is located at Sulaibiya in Kuwait in the *Rhanterium* steppe. The site location and details of the climate, topography, soil and vegetation have been described by Taha and Krueger (1981), and Halwagy *et al.* (1982). Five paddocks (1 × 2 km each) were allocated for the grazing study (Fig. 1). One paddock (No. 1) was assigned for continuous grazing, three paddocks (Nos 2–4) for rotational grazing and the fifth one for control. An area measuring 2.5 × 1 km was assigned as a holding pasture (Fig. 1). In order to study the grazing system, it was decided to select sheep as the main grazing herbivore since it is the most abundant animal in Kuwait's rangelands.

First year study

A flock of 129 mature ewes of Naeemi, Nigdi and Erbi breeds (43 from each breed) were used for initiation of a study in June 1981. The ewes were purchased from local sheep breeders after being examined for specific breed characteristics, deformity, disease and external injuries. Another flock of 43 ewe-lambs (6–7 months old) were also purchased for future grazing studies. They were kept under close observation in the holding pasture (Fig. 1), identified by ear tags and vaccinated against infectious diseases. The entire study was divided into two periods. During the first period (first year) grazing in the holding pasture was allowed, with provision for supplemental feed for a period of 243 days from July 1981 to February 1982. In the second period (first year) grazing of ewes was permitted under either the continuous or rotational grazing systems (Fig. 1) for a period of 122 days (March to June 1982).

Grazing in the holding pasture: The holding pasture was used for grazing of sheep on standing vegetation during the summer. The animals were allowed to graze during the morning for 5 hr and in the afternoon for 3 hr.

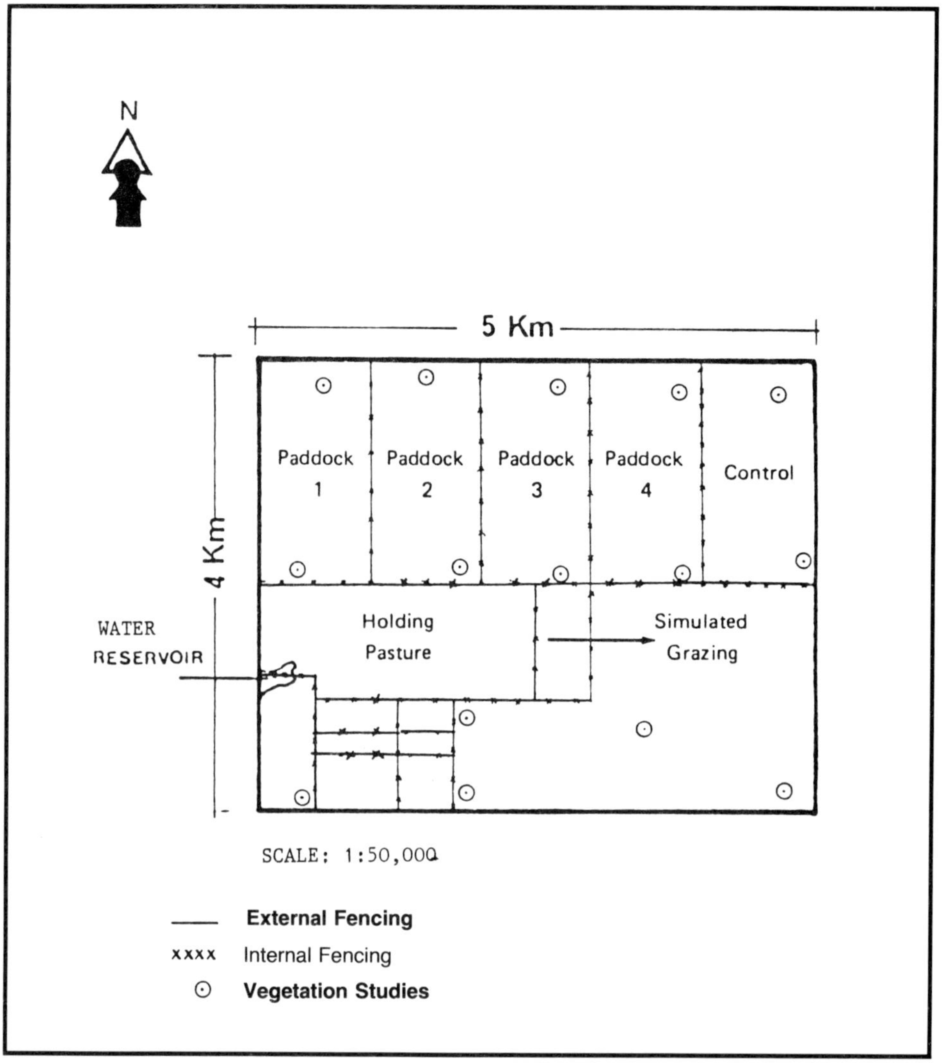

Fig. 1. Experimental grazing paddocks at KISR Sulaibiya site.

Supplemental feed was provided during the grazing intervals at noon and night depending on the need and performance of the sheep. The supplement feed comprised concentrate mixture and alfalfa (500–750 g/h/d) fed as and where required.

Breeding and lambing: Ewes of Naeemi and Nigdi breeds were bred in September and October while the Erbi ewes were bred during October and November. The new-born lambs were identified by ear-tags and their birth weight recorded.

Disease control measure: Routine disease control measures were followed, i.e. vaccination against infectious diseases and treatment for internal and external parasites.

Grazing studies; Two types of grazing systems, i.e. continuous and rotational, were used in the experimental pasture. A total of 156 ewes of Naeemi, Nigdi and Erbi breeds were used in grazing studies in the first year. A flock of 39 ewes comprising 13 each of Naeemi, Nigdi and Erbi were assigned to continuous grazing, and another flock of 117 ewes comprising 39 each of Naeemi, Nigdi and Erbi were assigned to rotational grazing allowing 5.1 ha/sheep for 122 days from March to June. Fresh water was available *ad libitum* in paddock sheds. It may be noted that 156 suckling young lambs were included with above mature ewes during the grazing studies.

Nutritional evaluation of major range plants was carried out through different phenological stages using the method of AOAC (1975) and Mehrez and Orskov (1977).

Second year study

Grazing in the holding pasture: The experimental flock comprised ewes and lambs. After withdrawing the sheep from the grazing treatments, they were placed in the holding pasture, commencing July 1982. The amount of supplement feed provided to the sheep was 1 kg/h/d in addition to grazing, and was maintained at that level for two months during the flushing time. During the months from October to November the sheep were provided with 750 g/h/d supplement feed which was later reduced to 500 g feed h/d for two weeks in December. Thereafter, the supplement feed was withdrawn and the animals completely depended on grazing.

Breeding and lambing: Ewes of Naeemi, Nigdi and Erbi breeds were bred during August to October after flushing for a period of 45 days. Lambing started during late December and continued for two months. The management of ewes and lambs during lambing time was similar to that of the first year.

Grazing studies; A flock of 232 sheep were used for the seasonal grazing studies. The flock comprised 172 ewes and 60 yearling female lambs (lamb crop of 1981–1982). A total of 58 sheep were used in a continuous grazing system comprising 16 each of Naeemi and Nigdi (11 ewes + 5 yearling lambs) and 26 Erbi (21 ewes + 5 yearling lambs). The flock in the rotational grazing system comprised 48 each of Naeemi and Nigdi ewes (33 ewes and 15 yearling lambs) and 78 Erbi ewes (63 ewes + 15 yearling lambs). The total grazing period during the second year was extended to 195 days from mid-December 1981 to the end of June 1982. The sheep depended mainly on the range plants and stocking rate was 3.45 ha/sheep. It is worth noting that each mature ewe (except yearlings) of the experimental flock had a suckling young lamb.

At the end of the second year's grazing season in June, four young lambs (5–6 month old) from each breed from both grazing systems were sacrificed for evaluation of dressing percentage. The live weight changes of ewes and lambs were monitored once every two weeks throughout the two year study period. The lambing percentage and birth weight of lambs of the three breeds were compared.

RESULTS AND DISCUSSION

Description of sheep breeds; The three breeds of sheep which were selected for grazing studies are described briefly for identification as follows:

Naeemi sheep: Naeemi sheep are characterized by a white body colour and white legs, sometimes with brown spots. The colour of the head may be brown or black. The fat tail is fairly large and compact. The general view of the animals shows that the body is fairly uniform and compact. The legs are shorter than those of either Erbi or Nigdi sheep. Naeemi sheep are meat-producing animals and produce wool of fairly good quality. They appear to be able to withstand harsh climate and poor vegetation. The mean live weights of Naeemi rams and ewes are 61.6 and 42.6 kg respectively.

Nigdi sheep: Nigdi sheep are medium-sized animals having a black body coat and white face and legs. They may occasionally have a completely white body coat. Their body and legs appear long and the tail is fat and long. Average mature live weight of the Nigdi ram and ewe are 60.5 and 38.2 kg, respectively. These animals are also a good meat-producing breed.

Erbi sheep: Erbi sheep have a black body coat with a fairly large black head and a distinct curved nose. They may occasionally be brown to black in colour. The body looks fairly uniform and the tail is fat and elongated. Erbi sheep inhabit flat areas of the south and southeast of Iraq and Kuwait. The mean mature live weight of the ram and ewe are 67.8 and 47.0 kg, respectively. Erbi sheep are known to be good for meat production and the breed survives extremely well on aridlands, as they are able to walk long distances in search of good pasture.

Published information regarding the characteristics of different breeds of Arabian sheep is lacking. A brief description and the production performance of Awassi sheep have been given by Williamson and Payne (1978), but further investigations are required to study the characteristics, origin, utility and meat production potential of major local breeds of sheep. The future selection of a suitable breed for the improvement of meat production in this part of the world is important.

Range plants and their quality: The main range plants grazed by the experimental flock comprised *Rhanterium epapposum*, *Plantago* spp., *Cutandia memphitica*, *Koelpinia linearis*, *Cyperus conglomeratus*, *Panicum turgidum*, *Stipagrostis plumosa*, *Gynandriris sisyrinchium*, *Ononis serrata*, etc. When first introduced into the grazing land, the animals were allowed to initially graze on standing mature range plants from July and grazing continued throughout the year. The average crude protein in the major shrub *Rhanterium epapposum* ranged from 40–80 g/kg dry matter during maturity. The mean crude protein in the mature forbs and grasses were 46.4 and 34.8 g/kg respectively. During the winter and spring the vegetation quality appreciably improved. The mean crude protein value of *Rhanterium epapposum* during these seasons was 230 and 110 g/kg dry matter respectively. The annual grasses and forbs contained crude protein values of 96 and 182 g/kg respectively. Along with the maturity of the plant, especially between March and April, there appeared to be a drastic drop in the crude protein content and an increase in the fibrous components. The plants grazed by the animals during the long summer months and early autumn provided poor nutrition. Provision of supplemental feed was made in the summer, autumn and part of the winter depending on rainfall and vegetation growth.

Animal performance during the first year (1981–1982)

Performance of ewes in the holding pasture: The results of animal response with respect to live weight changes of three breeds of ewes during a period of 365 days are shown in Table 1. The sheep were allowed to graze in an extensive system and the study period was divided to six experimental units. Table 1 results showed that all three breeds of Arabian sheep had lost live weight during the period from July to September when they were exposed to harsh climate and were grazing on standing mature range plants. Nigdi ewes lost significantly less live weight than the other breeds. The animals responded well to supplemental feeding during September to November. Withdrawal of supplement feeding in late November to early January resulted in a minor live weight loss in all ewes ranging from 11 to 50 g/h/d. Although, the quality and quantity of range vegetation improved in January and February, supplement feeding was given in order to support the additional nutritional needs of the animals during late pregnancy, and therefore all animals showed gain in live weight. Naeemi and Nigdi ewes gained significantly more than the Erbi ewes. At that time the Naeemi and Nigdi ewes were in late pregnancy, and the Erbi ewes were in early pregnancy, whereas the Erbi lambs were in the growing stage.

The loss of live weight in March–April which has been observed ir Naeemi and Nigdi breeds was mainly due to lambing. The Erbi ewes were still in the late stage of pregnancy at that time.

Table 1. Live weight changes (g/d) of three breeds of Arabian sheep under rangeland conditions of Kuwait (from July 1981 to June 1982).

Experimental period	*Dietary provision*	*Naeemi ewes*	*Nigdi ewes*	*Erbi ewes*
July to Sep. 1981 (80 days)	Grazing	-85.6^{a}	-25.6^{b}	-88.5^{a}
Sep. to Nov. 1981 (60 days)	Grazing + supplement feeding*	$+87.3^{a}$	$+74.3^{a}$	$+84.0^{a}$
Nov. to Jan. 1981–1982 (60 days)	Grazing	-50.0^{a}	-11.0^{b}	-46.5^{a}
Jan. to Feb. 1982 (37 days)	Grazing + supplement feeding*	$+179.5^{a}$	$+178.6^{b}$	$+135.9^{a}$
Mar. to April 1982 (61 days)	Grazing	-54.6^{a}	-93.9^{b}	$+110.0^{c}$
May to June 1982 (61 days)	Grazing	-120.8^{a}	-100.4^{a}	-163.4^{b}

a, b, c in the same row indicate means which are significantly different at $P \leq 0.05$.

Table 2. Some reproductive and productive parameters of Naeemi, Nigdi and Erbi sheep as recorded in Kuwait rangelands 1981–1982.

	Breed of Sheep		
Parameters	*Naeemi*	*Nigdi*	*Erbi*
Number of ewes	43	43	43
Number of ewes conceived	39	41	28
Conception rate (%)	91	95	65
Number of ewes aborted	1	2	3
Number of ewes lambed	38	39	25
Number of lambs born alive	40	39	25
Number of twin births	2	0	0
Lambing (%)	93	91	58
Lamb survival up to 3 months of age (numbers)	40	38	25
Number of ewe mortality	1	1	1
Birth weight of lambs (kg)	4.22	4.03	4.81
Wool production (g/sheep)	1582	—	1648

— = not determined.

Reproductive performance: The results of the reproductive performance of the three breeds of sheep are shown in Table 2. It should be noted that the Erbi ewes were mated later than the other two breeds. Moreover, the ram : ewe ratio was also different (Table 2). Unavailability of pure breed Erbi rams during the first year was the main reason for using the differential ratio of ram : ewe. The results show that fertility, as judged from the lambing percentage, was higher in Naeemi and Nigdi. The results show that the newborn lambs of the Erbi were significantly ($P \leq 0.05$) heavier than those of either the Naeemi or Nigdi breeds. There was no significant difference ($P \leq 0.05$) between the birth weight of Naeemi and Nigdi lambs. Research findings on the birth weight of lambs of different breeds of Arabian fat-tailed sheep are limited. Ampy and Rottensten (1968) observed that the birth weight of lambs of the Awassi breed ranged from 3.9 to 4.4 kg (mean 4.15 kg) in both sexes. Our studies show a birth weight of 4.22 kg for Naeemi lambs (Awassi origin), which is consistent with Ampy and Rottensten (1968). Our present studies clearly show that Arabian fat-tailed sheep produced heavier lambs than the Merino breed. Cartright and Thwaites (1976) reported an average birth weight for Merino lambs of 3.1 kg.

Animal response to grazing systems (1981–1982): Responses of the three breeds of local sheep (Naeemi, Nigdi and Erbi) were monitored in the continuous and rotational grazing systems from March to June, a period of 122 days. The main parameter of the study was the live weight changes of the young lambs and the mature ewes.

Young lambs: The mean live weight changes in lambs (suckling and grazing) are shown in Table 3. The lambs were born at different dates beginning from February so the rate of body weight gain of individual lambs differed appreciably. The suckling and grazing lambs of all three breeds showed a

Table 3. Daily change (g) of live weight of young and mature ewes in two types of grazing systems (1981–1982).

Breed	*Type*	*Continuous grazing*	*Rotational grazing*
Naeemi	Young lambs	141	133
	Ewes	−87	−91
Nigdi	Young lambs	157	134
	Ewes	−88	−93
Erbi	Young lambs	190	140
	Ewes	−44	−50

There is no significant difference between treatment means of the two grazing systems with respect to daily live weight changes.

sharp increase in live weight from March to the end of April in both continuous and rotational grazing systems. The daily gain in Naeemi and Nigdi lambs was up to 380 g/h. During this time, the quality of range vegetation was good in terms of nutrient content. The lambs depended on their dams' milk through suckling and started to graze some range forage. The difference between the live weight gain in continuous and rotational grazing systems was not significant ($P \leq 0.05$).

Mature ewes: It should be mentioned here that, during initiation of the grazing trial in March, Naeemi and Nigdi ewes were either in a late stage of pregnancy or had just started lambing. The loss of live weight of Naeemi and Nigdi ewes was mainly associated with parturition during March–April 1982. Live weight losses during May and June (Table 3) were attributed to deteriorated range plant quality. The results also show that live weight losses in Naeemi and Nigdi ewes during the last two months of the trial (May and June) were more drastic than those of all ewes earlier in the season (March and April). The live weight loss in Erbi ewes during the four month grazing trial was significantly ($P \leq 0.05$) less than that of either Naeemi or Nigdi ewes in both the continuous and rotational grazing systems.

When comparing the performance of all sheep breeds in terms of live weight changes in both types of grazing systems (Table 3), it becomes apparent that the two treatments did not differ significantly. A possible reason may be the drought conditions, which might have masked treatment effects.

Animal performance during the second year (1982–1983)

Performance of ewes and lambs in the holding pastures: The response of the three breeds of sheep during the summer grazing in the holding pasture with provision of supplement feeding is summarized in Table 4. The live weight gain during July–August in ewes ranged from 27 to 48 g/h/d. The live weight gain in Nigdi ewes during the first 49 days was significantly lower than that of the other two breeds. However, during the period from mid-August to mid-October the Nigdi ewes gained live weight, whereas the other two

Table 4. Mean live weight changes (g/d) of three breeds of Arabian sheep under the rangeland conditions of Kuwait (1982–1983).

Experimental period	*Dietary provision*	*Naeemi*		*Nigdi*		*Erbi*	
		Lambs	*Ewes*	*Lambs*	*Ewes*	*Lambs*	*Ewes*
July to mid-Aug. (49 days)	Grazing + supplement feeding	65	48	59	27	65	44
Mid-Aug. to mid-Oct. (60 days)	Grazing + supplement feeding	−52	−59	−2	25	−10	−77
Mid-Oct. to mid-Dec. (56 days)	Grazing + supplement feeding	18	53	15	14	38	−17
Entire 165 days	Grazing + supplement feeding	7^{b}	13^{d}	22^{a}	22^{c}	29^{a}	-21^{e}

a, b: mean values of lambs with different superscript are significantly different ($P \leq 0.05$).
c, d, e: mean values of ewes with different superscript are significantly different ($P \leq 0.05$).

breeds lost live weight appreciably, although all the animals were subjected to the same treatment, i.e. grazing and supplement feeding. The overall daily gain of Nigdi ewes during the 165 day study period was significantly better than in the other two breeds. The Erbi ewes lost a mean live weight of 21 g/h/d but the young lambs of Erbi breed showed appreciably higher live weight gains than the other two breeds.

The adverse environmental conditions associated with poor vegetation quality during the summer and autumn were reflected in the poor performance of both lambs and ewes.

Table 5. Reproductive and productive parameters of three breeds of Arabian sheep for the year 1982–1983.

Parameters	*Breed of Sheep*		
	Naeemi	*Nigdi*	*Erbi*
Number of ewes	42	32	80
Number of ewes conceived	35	31	65
Number of ewes aborted	1	0	3
Number of ewes lambed	34	31	57
Number of lambs born alive	34	31	63
Number of twin births	0	0	3
Lambing (%)	81	96.9	78.8
Birth weight of lamb (kg)	4.38	4.26	4.35
Lamb survival up to 3 months of age (numbers)	32	28	59
Number of ewe mortality	1	1	1
Wool production (g)	2185	—	2168

— = not determined.

Reproductive performance: The results of the reproductive performance of the three breeds of sheep is summarized in Table 5. The data show a drop in the lambing percentage in Naeemi breed, whereas an improvement was noted in the Nigdi and Erbi breeds in the second year compared to the first year. In the second year (1982–1983) there was no significant difference between the birth weight of lambs due to breed differences. The survival percentage of lambs up to 3 months of age in Naeemi, Nigdi and Erbi breeds was 94, 90 and 94 respectively. An improvement in wool production in both Naeemi and Erbi was observed compared to the first year results.

Animal response to grazing system (1982–1983)

Young lambs: The results of live weight changes of young lambs and ewes are shown in Table 6. There was no significant difference between grazing systems with respect to lamb live weight change during the 195 day study period. However, an apparent better performance of lambs was obtained under the continuous grazing system. During each part of the study from birth to weaning age of the lambs the rates of daily live weight gain ranged from 283 to 329 g/h irrespective of breed and grazing system. During this period and up to the end of March the quality of range vegetation was very good. The lambs continued to gain live weight; total live weight of young lambs during the 4th week of May in Naeemi, Nigdi and Erbi breeds was 34, 31 and 33 g/h/d, respectively. Thereafter, young lambs showed some losses in live weight ranging from 14 to 91 g/h/d.

Ewes: The mature ewes of Naeemi, Nigdi and Erbi breeds have performed significantly better in the continuous grazing system than in the rotational grazing system (Table 6).

Yearling lambs: The results of animal performance of yearling lambs (lamb crop of 1982) are summarized in Fig. 2. At the beginning of the grazing season (mid-December), the lambs did not gain live weight for a period of two weeks. During the month of December 1982, the range vegetation started to grow and supplied the maintenance needs of the grazing sheep.

Table 6. Daily live weight change (g) of young lambs and mature ewes in two types of grazing systems (195 day study period 1982–1983).

Parameter	*Type*	*Continuous grazing*	*Rotational grazing*
Naeemi	Young lambs	137	118
	Ewes	24^{a}	15^{b}
Nigdi	Young lambs	145	126
	Ewes	12^{a}	-5^{b}
Erbi	Young lambs	141	126
	Ewes	24^{a}	5^{b}

a, b = means in the same row with different superscript differed significantly ($P \leq 0.05$).

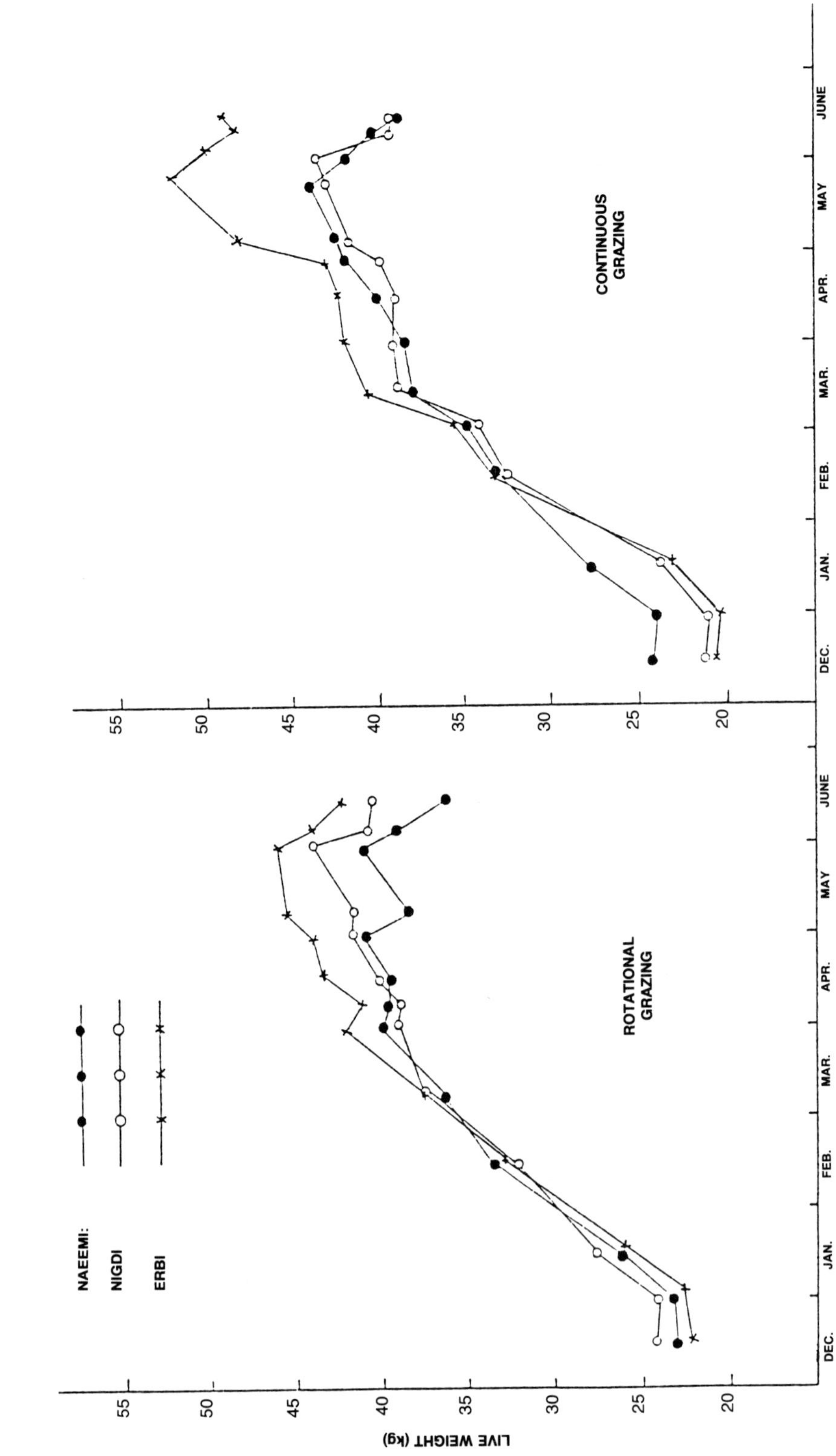

Fig. 2. Live weight changes of three breeds of yearling lambs in two types of grazing systems (1982–1983 studies).

Significant positive responses of lambs to grazing were realized from early January to the end of March for all the three breeds of sheep under both types of grazing systems. The daily gain in the rotational grazing system was 218, 199 and 255 g in Naeemi, Nigdi and Erbi yearling lambs, respectively, and the corresponding values for the continuous grazing system were 217, 185 and 295 g respectively. The rate of live-weight gain during the period from April to May was greatly reduced and followed a similar pattern as observed with younger lambs. The differences between the two grazing systems were not significant. However, the Erbi yearling lambs appeared to out-perform either the Naeemi or Nigdi sheep during the study period. In June, the animals were not able to maintain their live weight and losses ranging from 103 to 123 g/d were recorded. The losses in Erbi sheep were considerably lower than in the other two breeds.

Dressing percentage: Table 7 shows the dressing percentage of three breeds of lambs which were slaughtered at the end of the grazing trial (1982–1983). Similar to the ewes, the young lambs which grazed in June lost live weight considerably prior to being slaughtered as a result of environmental and nutritional stresses imposed upon them. There were no significant differences between breed or grazing system with respect to dressing percentages except for the fat tail of Erbi lambs which was heavier than that of either Naeemi or Nigdi lambs.

Table 7. Carcass and some non-carcass components of three breeds of lambs; values are means of four lambs.

	Naeemi	*Nigdi*	*Erbi*
Live weight (kg)	30.55	29.34	32.73
Carcass weight (kg)	10.58	10.13	11.55
Dressing (%)	34.57	34.36	35.18
Weight of kidney (g)	79.73	79.75	79.23
Weight of kidney fat (g)	58.73	53.20	47.35
Weight of fat tail (g)	725	762	1075

LITERATURE CITED

Amir, D. and R. Volcani. 1965. The sexual season of the Awassi fat-tailed ewe. *J. Agric. Sci. (Cambridge, UK)* **64:** 83–85.

Ampy, F. R. and K. U. Rottensten. 1968. Fertility in Awassi sheep: 1 – Seasonal influence on fertility. *Trop. Agric. (Trinidad)* **45:** 191–197.

AOAC. 1975. *Official methods of analysis,* 12th ed. Washington D.C.: Association of Official Agricultural Chemists.

Cartright, C. A. and C. J. Thwaites. 1976. Influence of maternal nutrition and high temperature on the growth and proportions of merino foetuses. *J. Agric. Sci. (Cambridge, UK)* **86:** 576–580.

Demiruren, A. S. 1972. Sheep production potential in the eastern Mediterranean. *World Animal Review* **2:** 21–27.

Halwagy, R., A. F. Moustafa and Susan M. Kamel. 1982. On the ecology of the desert vegetation in Kuwait. *J. Arid Environ.* **5:** 95–107.

Johnson, H. D. and G. L. Hahn. 1982. Climate and animal productivity, vol. 2. Edited by M. Recheigh, Boca Raton, Florida.

Mehrez, A. Z. and E. R. Orskov. 1977. A study of artificial fibre bag technique for determining the digestability of feeds in the rumen. *J. Agric. Sci. (Cambridge, UK)* **88:** 645–650.

Mittal, J. P. and P. K. Ghosh. 1980. Desert sheep withstand salinity and a scarcity of water. *Indian Farming* **30:** 31–33.

Payne, W. J. A. 1969. Problems of the nutrition of ruminants in the tropics. *In:* D. Cuthbertson (ed), *Nutrition of animals of agricultural importance,* Part 2. Pergamon Press, Oxford.

Saleh, E., E. Galal and Y. S. Ghanem. 1970. The interaction between breeds of sheep and pattern of production and management. *U.A.R. J. Animal Prod.* **10:** 7–12.

Taha, F. K. and W. C. Krueger. 1981. Kuwait's range research model for development of a viable sheep industry. *In:* J. Menasse (ed), *Advances in food producing systems for arid and semi-arid lands.* Academic Press, New York.

Williamson, G. and W. C. A. Payne. 1978. *An introduction to animal husbandry in the tropics.* Longmans, London.

Younis, A. A., F. Al-Mahmoud, E. A. El-Tawil and A. S. El-Shoboksy. 1977. Performance and heat tolerance of Awassi lambs as affected by early shearing. *J. Agric. Sci. (Cambridge, UK)* **89:** 565–570.

Animal range relations on arid and semi-arid rangelands

WILLIAM C. KRUEGER

Department of Rangeland Resources, Oregon State University, Corvallis, Oregon 97331, USA

ABSTRACT

Successful production of livestock from rangelands results from an integration of plant and animal ecology within a practical management framework. Research and experience have laid the foundation to understand ruminant nutrition including palatability, nutrient utilization and nutrient requirements of livestock. When this is incorporated with knowledge of nutrient intake, adequacy of the range to meet production needs, as well as habitat utilization and grazing impacts on a site, an animal management model can be developed. In the arid and semi-arid areas of North Africa and the Middle East, much basic research remains before the potential of these rangelands to contribute to animal production is realized. Control of livestock grazing remains a central social problem that must be solved if animal production is to be improved. If workable range livestock production programs are to be developed, the local, site-specific scientific production problems must be solved along with the social problems of management.

INTRODUCTION

The principal agricultural use of arid and semi-arid rangelands is production of livestock. Throughout the world, families living in these sparsely populated areas depend heavily on livestock products for maintenance of their way of life or even survival. Modern pressures of economics and increasing population densities demand greater efficiency of production as the basic productivity of the land in many areas slowly decreases. Livestock production practices that deteriorate the rangelands are the rule and consequently both the land and the people dependent upon the land suffer. Eventually, livestock production from arid and semi-arid lands will cease because the resource will have been seriously depleted, unless livestock production practices stem from ecological sciences and are based on a principle of sustainable yield. If livestock production on these rangelands is developed from science, these enterprises can be profitable in both economic and social aspects. Additionally, benefits from healthy ecosystems will accrue to the cultures and societies that are tied to the rangeland areas. The purpose of this paper is to provide, in general terms, the results of research and experience that can be used to integrate plant and animal ecology within a practical management framework.

PALATABILITY

Rangeland livestock are of necessity managed differently than livestock on irrigated pastures. As livestock encounter the array of vegetation in any range site, they will select specific plants or parts of plants over others. The range in relative preference among animal species, is extensive and varies considerably over seasons and with respect to past and current management practices. Palatability of vegetation is defined by the animal species through their special senses of sight, smell, taste and touch (Krueger *et al.* 1974). Generally, livestock taste a plant for primary discrimination and the other senses are supplementary. Preference for forages is a learned behaviour and varies with the condition of the livestock. As hunger increases, selectivity declines. A grazing animal is integrating a multitude of stimuli as it moves across the range, each being sensed and sorted according to the changing animal condition and diurnal fluctuations. As a result, preference is an extremely complex behaviour that has not yet been satisfactorily modeled.

Substantial research has been conducted in an attempt to define the non-animal components of palatability (Heady 1964). Palatability is a changing phenomenon that results from the chemical composition of forages, toxins, morphological characteristics, growth stage and succulence. Of the external factors that can predict palatability of a forage, the plant community is the most powerful. When an animal makes a choice, the options for selection among forage species is a principal determinant of what the choice will be. Sheep, camels and cattle would face the same array of vegetation in the community and because of the varying animal influences, the choice of diet would be different. When one of these animal species encounters the same plant species in different plant communities, the relative preference for that plant would be different for each community.

Measurement of palatability of rangeland forages has been approached from many different aspects (Holechek *et al.* 1982). Generally, utilization of forage or fecal analysis is used to define animal diets.

Once the diet of animals for a given range can be defined, the area is more amenable to manipulation to reach specific production objectives. Management practices can be designed to minimize relative abundance of unpalatable species or to change the mixture of species to improve palatability of the dominant vegetation. Success is dependent upon understanding of synecology, i.e. the ecological potential of a site and its limitations.

As the range management program is developed to direct plant succession, the goal of management should be to understand and focus the impacts these sucessional changes will have on palatability of vegetation as the mixture of plant species changes through time. Moreover, when we understand why animals are eating or are not eating vegetation, we are then in a position to do something about it to achieve desired objectives. Simply put, this insight allows the manager to select specific actions to solve specific problems.

NUTRITION OF RANGE LIVESTOCK

Nutrition of livestock has been studied throughout the world for over a hundred years. The role and availability of macronutrients from rangelands are well documented (Cook 1970). Micronutrient availability is highly variable among range sites and generally micronutrients needs are best met through supplementation, usually with a trace mineralized salt.

As range vegetation matures and dries, lignin, cellulose, and calcium increase in concentration while water, soluble carbohydrates, protein, carotene and most minerals decline. This results in substantial changes in the ability of the vegetation to meet nutritional needs of livestock. Grasses undergo more dramatic fluctuations than shrubs which are more stable over time. When vegetation matures, grasses are a relatively good source of energy while shrubs are usually deficient in energy but provide a nearly adequate supply of protein, carotene and phosphorous. The leaves of most plants are more nutritious than the stems which, by their nature, are highly fibrous.

Normal seasonal trends are well defined for grasses and shrubs, recognizing that certain species are variants from the normal situation. The decline in quality with maturity can be arrested and directed by timely grazing. Grazing a plant defers maturity and in arid lands once a plant is grazed in an early growth stage there is seldom sufficient time to allow the plant to reach full maturity. When this occurs, the vegetation remaining at the end of the growing season cures in a higher nutritional status. Balancing the loss of vigour in plants due to stress of grazing and the value of higher quality feed in the dormant period will result in more efficient production of livestock. The loss in yield can be more than compensated by the increase in quality of residual herbage.

When grazing is excessive, instead of improving, the quality of residual herbage for the dormant period of that range will deteriorate. The more palatable and usually most nutritious forages are removed and thus their intake decreases. Consequently, even if the less palatable plants are as nutritious as the palatable plants, total nutrient intake decreases.

The nutrients of greatest concern for production of range livestock are protein, energy, water, minerals (especially phosphorous) and vitamin A; since these nutrients are often limiting on rangelands. Protein is essential for growth and animals need a continuous supply. If intake is adequate, 4.4% digestible protein is required for maintenance and 5.4% is required during lactation. Protein is often satisfactory in shrubs but is usually deficient in mature grass.

Energy is necessary for most physiological reactions of the animal and, like protein, is needed in a continuous fashion. If intake is adequate, livestock should receive 46% total digestible nutrients (TDN) or 1.83 megacalories of digestible energy per kilogram of forage consumed as a maintenance ration and 56% TDN or 2.47 megacalories of digestible energy per kilogram of forage consumed during lactation.

Mineral deficiencies or toxicities are common on rangelands but usually specific site factors dictate mineral nutrition status. Phosphorus is generally deficient in arid and semi-arid range and must be applied at a level of 0.17% of the diet for maintenance and 0.22% during lactation. This can be achieved by grazing of shrubs in many situations or by supplementation.

Growing animals grazing mature grass rangelands will show vitamin A deficiency if held on the mature grass for 60 days or 120 days for mature livestock. Since livestock can store vitamin A for 60 days, the normal practice in the United States is to supplement vitamin A after this period.

Water is critically important to livestock production on arid and semi-arid rangelands. Livestock acquire water from succulence in vegetation, dew, free water sources and their own metabolic activity. The need for water varies with kind and breed of livestock, temperature, age of the animal, level of activity, production function (e.g. during lactation water needs increase), frequency of watering and other factors. If water is deficient, food consumption decreases and animal growth slows.

MANAGEMENT

The adequacy of rangelands to meet livestock production requirements varies considerably (Vavra and Raleigh 1976). Rangelands dominated by annual vegetation are usually adequate in most nutrients when they dry in early summer. To maintain adequate levels of livestock performance, supplementation is required especially for protein and phoshorous. If intake is restricted in any way, an energy supplement will also be required. If rangelands are dominated by perennial grasses, they are nutritionally adequate during the growing season. If grazing is deferred until the forage is mature, they are deficient in protein, phosphorous and vitamin A. If grazed for a short period immediately prior to seed formation, perennial grasses will often cure to be adequate in all the macronutrients. When shrubs mature, the range is generally deficient in energy and if it is not supplied by grasses on the range, energy will need to be provided in a supplement. By carefully managing the existing forage base, the need and expense of supplementation can be minimized.

A variety of grazing systems have been developed for use on arid and semi-arid rangelands. These include various rotational systems that regulate animal use. For the most part, grazing systems have been developed to improve rangelands, not to enhance animal production. Proper implementation of a technical grazing system should influence plant succession and direct the community to a higher seral status. However, this is often accompanied by a decline in animal production.

It is possible to develop grazing programs to enhance both the productivity of the range and the livestock. This is done by understanding the growth requirements of the plants and animals on each site and during each growth stage for both range forage and range livestock. Animal use must be regulated to maintain or improve the range, but at the same time consideration of animal needs for food, water and environment must be made and these needs accommodated.

The most difficult problem for range managers in North Africa and the Middle East is to actually regulate livestock use. The long traditions of grazing in common and, in many areas, large populations of economically marginal farmers dictate against restrictions on grazing. The solution to this problem is rooted in sociology and it must be solved if animal production is to be improved. Without control of livestock grazing, there will be no progress in range management of lasting value to the people of this part of the world. Solution to the social problems should accompany the research that seeks to understand ecology and agricultural production.

Specific answers for the arid and semi-arid rangelands of the Middle East are developing but many details of extreme importance are lacking. For the present, range management for livestock production is still largely an art that is supplemented by science. The goal in this part of the world should be to reverse the emphasis of art and science so that range management will be largely a science and as a consequence more productive and more predictable.

LITERATURE CITED

Cook, C. W. 1970. Energy budget of the range and range livestock. *Colo. Agr. Exp. Sta.* Bull. TB 109.

Heady, H. F. 1964. Palatability of herbage and animal preference. *J. Range Manage.* **17:** 76–82.

Holechek, J. L., M. Vavra and R. D. Pieper. 1982. Botanical composition determination of range herbivore diets: a review. *J. Range Manage.* **35:** 309–315.

Krueger, W. C., W. A. Laycock and D. A. Price. 1974. Relationships of taste, smell, sight and touch to forage selection. *J. Range Manage.* **27:** 358–362.

Vavra, M. and R. J. Raleigh. 1976. Coordinating beef cattle management with the range forage resource. *J. Range Manage.* **29:** 449–452.

Analysis of animal production systems in marginal lands (with a case study of The United Arab Emirates and experiences from Eastern Africa)

AHMED E. SIDAHMED*

FAO Expert, Range and Livestock Development Project, c/o UNDP, P.O. Box 358, Tripoli, Libya

ABSTRACT

Arid and semi-arid regions are characterized by erratic and very low rainfall patterns. Soil fertility is poor in most of these lands. The areas are marginal in their capabilities to produce food crops. However, for thousands of years, domesticated animals existed in arid ecosystems, usefully converting scanty primary production to animal protein. Traditional adjustments to fluctuating forage and water supplies were maintained by a flexible and cyclic movement pattern of the livestock and their owners (nomads, semi-nomads and transhumants).

In this paper, animal production systems in some East African (Sudan, Kenya and Ethiopia) and Arabian Gulf (United Arab Emirates) regions are analysed. Livestock distribution, structure and production variables are presented. Constraints to livestock production are discussed. They include exogenous elements (administrative and technological) as well as biological components (overgrazing, cultivation of grazing lands, unprescribed fires and irrational construction of water points. Range and animal-related operations and decisions suggested include: (1) reseeding of desirable plants, (2) sub-division and rest rotation, (3) supplementary feeding, (4) water development, (5) fodder cultivation, (6) fire management, (7) herd composition, (8) stocking rate, (9) disease control, and (10) offtake and marketing. Specific measures suggested to improve rangelands and feed supply include: (1) restriction of cultivation in marginal lands to fodder crops, (2) intercropping feed-food crops, (3) encouraging nitrogen-fixing indigenous leguminous trees, (4) prohibition of wood cutting in marginal lands, and (5) feed conservation. A case study of the animal production system in the United Arab Emirates has shown an increase in livestock and feed production in that country during the last 3 years. However, a remarkable decline in animal/human and cultivated land/human ratios suggests an increasing trend towards importing food products into that country.

*Present address: FAO, Room C-637, Via delle Terme di Caracalla, 00100 Rome, Italy.

INTRODUCTION

Marginal lands are characterized by questionable physical or economic capabilities for sustaining a specific use (Society for Range Management 1974). Semi-arid and arid regions are marginal in their capacity to provide cultivated food crops and are more suitable for extensive (pastoral) animal production systems. For thousands of years domesticated animals existed in arid ecosystems. The pastoral way of life (nomadic, semi-nomadic and transhumant) has resulted from the traditional adjustments to fluctuating forage and water supplies where only extreme mobility can maintain the flocks and herds (French 1966). Examples of traditional adjustments to overcome water shortages were discussed by Sandford (1983). Adjustment to the forage fluctuations coincides with those of water in many situations. Pastoral nomadism is a classic example of a flexible and opportunistic utilization of widely scattered resources (Johnson 1979). Livestock owners in pastoral systems depend on their animals for milk and meat and sometimes for exchange for money. It is therefore a misconcept to consider pastoralism in any context other than being a dynamic adjustment to a difficult and very delicate environmental situation. However, extreme pressures in recent times have made it impossible for pastoralists to practise their long-acquired survival strategies, or to have the flexibility to perform short-term tactics (Cossins 1983).

In this paper, pastoral livestock systems as influenced by exogenous and natural factors will be discussed. Some emphasis will be placed on livestock and range improvement. A case study of the animal production systems of the United Arab Emirates (UAE) will also be presented.

HUMAN SUPPORT CAPACITY OF MARGINAL LANDS

The arid and semi-arid regions comprise 30 to 50% of the earth's surface. However, only 15% of the global population inhabit these lands (Hjort 1976). According to Pratt and Gwynne (1977) the "human support capacity" of arid and semi-arid zones of East Africa varies from 7 persons/km^2 in the semi-arid regions to 1 person/2 km^2 in the very arid regions (Table 1). Pastoralists adapt to the erratic and very low rainfall patterns by adopting a flexible migratory pattern. In spite of the long drought which struck the African marginal rangelands since the early seventies, areas such as Turkana District in Kenya suffer from a high population density (3 persons/km^2). In Turkana, large numbers of the people live under famine relief while others have abandoned animal rearing to cultivation in development schemes (Stiles 1983). Such exogenous factors (famine relief funds, resettlement schemes inappropriately located in marginal lands, modern technology, roads, veterinary and human health care) have subjected the people who live in these areas to extreme economic and environmental pressures in recent years.

The other example to be discussed here in relation to human support capacity is the situation in the very arid oil-rich states of the Arabian Gulf.

Table 1. Animal production systems and human support capacity of arid and semi-arid regions.*

Agro-ecological zone Rainfall (mm/yr)	Type	*Animal production systems*	*Human support capacity (persons/km²)*
100	Very arid	Nomadic pastoralism	0.5
200			
300			
400	Arid	Nomadic pastoralism	2.0
500			
600			
700	Semi-arid	Semi-arid and transhumance	7.0

*Source: Hjort (1976), Pratt and Gwynne (1977).

One example – the United Arab Emirates – will be highlighted. In a succeeding section of this paper livestock production in UAE as influenced by oil production will be discussed. The population of UAE increased from 227,000 people in 1979 to 1,040,000 in 1981 and currently the UAE (60–100 mm annual rainfall) is inhabited by 13.3 persons/km^2 (FAO 1974, UAE Ministry of Agriculture and Fisheries 1981). However, the person-to-land ratio of the agriculturally active work force is only 1 person/7 km^2. This fact protrays the influence of non-agricultural factors on human dynamics. It is also very worthy to note that in these countries the standard of living is very high and in most cases comparable to that of people in the developed nations of north America and western Europe. According to Hallam (1983), meat and milk consumption in some developing countries is currently rising at the rate of 6–7% per year.

THE PASTORAL PRODUCTION SYSTEMS

The extensive systems of animal production are those in which animals graze over large areas. Because of the scarcity and inconsistency of rain, or due to prohibitive terrain (e.g. mountains) such areas are not suitable for cultivation. Livelihood in these marginal lands is dangerously insecure. Typically, pastorals own large numbers of stock and are concerned about the structure of their herds. For example, Baggara nomads of the Sudan maintain a herd composition of 70 to 75% female cows (Sidahmed and Koong 1984). This high breeding capacity is considered as an insurance for a rapid herd recovery after drought.

Three classes of pastoralism were defined by Hjort (1976) and Sandford (1983).

(i) *Nomadic pastoralism*

These systems depend mainly on livestock for subsistence. Supplemental farming may be practised. Also some of the family members may inject substantial income to the system from wage employment in nearby urban locations. This class of pastoralists keeps multiple species of livetock "as a response to probability of drought and to fully exploit the often mixed vegetation species" (Dyson-Hudson 1982). Depending on the radius of the communal grazing domain, one livestock species predominates. For example, in western Sudan the Baggara raise large numbers of cattle – a choice favoured by a diverse grazing domain stretching southward to the highland savanna regions with annual rainfall above 750 mm (Sidahmed and Koong 1984). On the other hand, the Kababish Arabs of the Sudan are mainly camel herders, but they also own sheep and goats. It is interesting to note that sheep are considered by the Kababish as more tolerant to water shortage than goats (Sandford 1983). This is reflected in the structure of their livestock which is largely composed of camels, sheep and goats, the latter falling at the bottom of the list in terms of numbers.

Nomadic pastorals are extremely mobile (Fig. 1). Generally, all family members move with the herd especially during the rainy season. In normal years they will all return to their *dars* (home ranges) during the dry season. Migration is usually in search of water, forage and mineral supplementation. Although dry-season movements are long-term survival strategies, wet-season movements are opportunistic tactics aiming at securing maximum conversion of growing vegetation to secondary production in the shortest possible time. Any observer of the nomadic skills will be overwhelmed by their exceptional talent for weather forecasting.

(ii) *Semi-nomadic pastoralism*

These systems differ from the nomadic systems in herd structure and in the area which is covered during migration. Semi-nomads migrate in the semi-arid regions only (Table 1) and concentrate on one species of livestock. While elders, women and children stay in one camp (*dar* or *village*) all the year round and become engaged in cropping and rearing of a few selected breeding stock, the majority of adult male labour move with the animals. In these systems, subsistence is provided by both livestock and cropping. However, social status and future insurance are achieved through the animal husbandry component.

(iii) *Transhumance*

Transhumants practise farming and livestock husbandry. However, cropping is more important (in terms of subsistence or wealth) than animal rearing. In normal years, livestock movement occurs during the dry season and for short distances. For example, the Nilotic transhumant tribes of the Sudan move for short distances, mainly east or west of the river according to flooding. In some transhumant communities, livestock herders are hired from other ethnic groups. Movement of livestock may be undertaken to keep livestock away from crops during the growing periods (Sandford 1983).

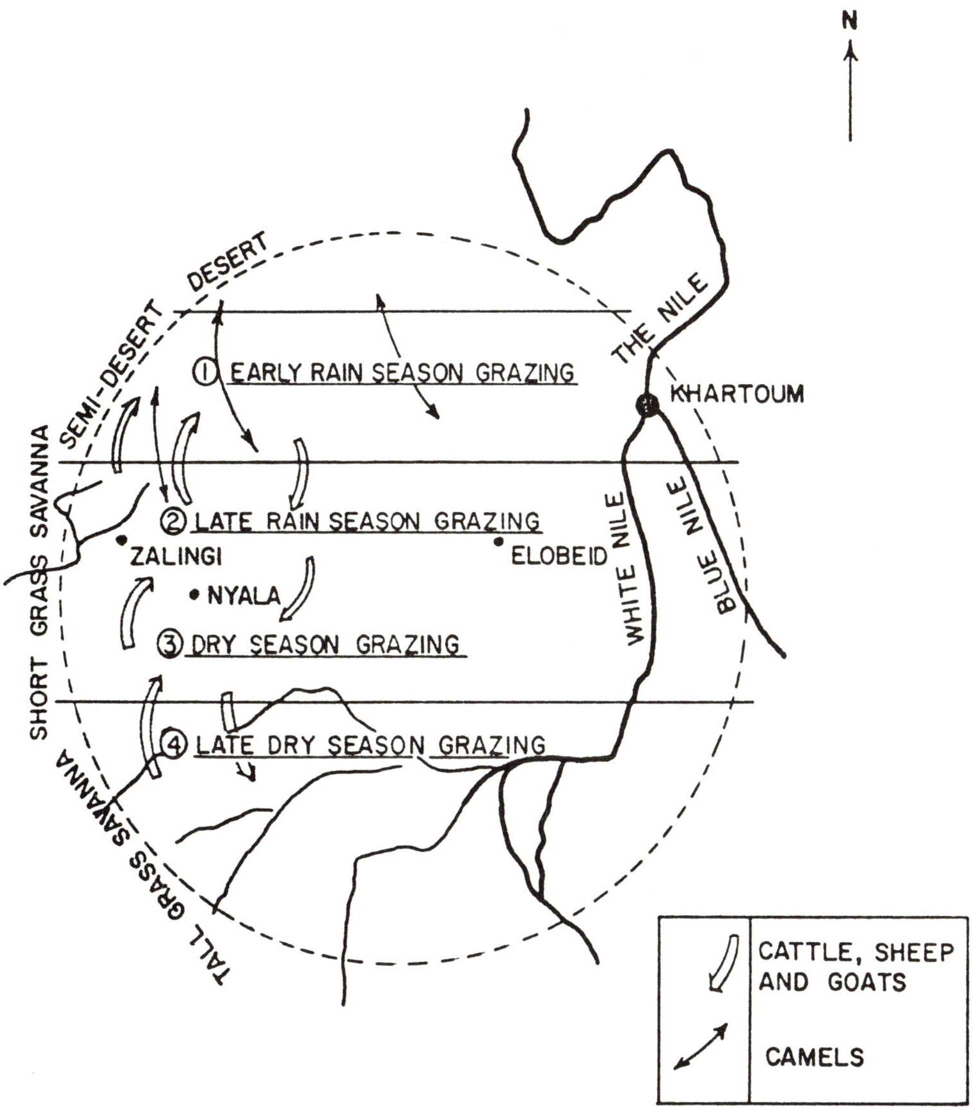

Fig. 1. Nomadic livestock movements in western Sudan. Source: Sidahlmed and Koong (1984).

THE DISTRIBUTION OF RUMINANTS IN MARGINAL ENVIRONMENTS

Hebivores can convert low-quality fibrous feeds to high quality products. This is particularly important where net primary production is limiting due to uncertain weather conditions. In such areas, cultivation is not reliable and livestock raising is considered the only feasible enterprise. The traditional strategies of pastoral societies are reflected in the cyclic migration patterns and in livestock species composition. Camels are more adapted to utilize desert shrubs and scrub and can be located up to 80 km from water (Lundholm 1976). They can endure 9 to 10 days between waterings.

The small ruminants (sheep and goats) are well adapted to the fluctuating environment of the arid and semi-arid regions. According to Winrock International (1983), small size, early maturity and low capital investment per head make small ruminants suitable to the needs of the limited-resource producers. Both species can select from a wide range of feeds extending from high-quality grass to bitter and tanniferous browse. The foraging behaviour of goats is particularly interesting. This species of livestock is a generalist opportunist and can exhibit a highly adaptive selection performance if resources are abundant (Sidahmed and Morris 1982). When the biomass is limiting, goats are generalists and select their diet from the most dominant species. Sheep and goats are complimentary to each other in their feeding strategies and also to cattle in mixed herds of ruminants (Winrock International 1983). They therefore contribute a flexibility to the pastoralists. Pastorals include sheep and goats in their herds as an insurance against disease or disaster, as a tool to trade, a reserve of ready cash, an easily expandable bank account and as a source of readily obtained food and fiber. The fact that both goats and sheep can travel without water for four to five days greatly increases their feeding range.

Cattle distribution in arid and semi-arid regions of Africa depends on the distribution of the tsetse fly and the trypanosome parasite. Local cattle are not well adapted to the climatic fluctuations in arid regions. Zebu cattle have a very slow rate of maturity and cannot tolerate thirst for more than one or two days. However, the predominance of cattle in the pastoral animal husbandry systems is attributed by Lundholm (1976) to high meat production and social considerations (e.g. prestige and dowry payment).

CONSTRAINTS TO PASTORAL ANIMAL PRODUCTION SYSTEMS

(a) *Administrative*

The pastoral animal production systems are subject to three series of constraints: normal (mostly seasonal, rainy versus dry periods), disastrous (epidemic diseases, fires) and irreversible long-term changes (Cossins 1983). A classical example of the latter is the loss of dry-season grazing of the semi-arid regions of the Sudan to cultivation. That country witnessed increased use of farm machinery during the last 30 years. Bush clearance was easier and cheaper in the marginal lands in the north (Daley Mechanized Crop Production Schemes) than in the densely forested south. This exploitation problem was aggravated by the establishment of deep-bore drinking water points in the desert fringes of western Sudan. However, all the water points are now abandoned because of rangeland depletion and by 1973 only 18% of the Daley Dry Farming Scheme was under cultivation. Historically, pastoralism is considered a strategic adaptation to natural constraints.

Consumption of milk or communal ownership of land are long-term pastoral strategies, while delaying the sale of males until maturity is a tactical decision. However, pastoral societies have come under unprecedented political, economic and environmental pressure in recent years (Pendelton and Van Dyne 1980). The strategic mobility and communal land use systems of the pastoral people have been tremendously restricted by administrative and institutional pressures. According to Dyson-Hudson (1982):

> "Pastoral populations are no longer allowed to adjust naturally to phenomena such as drought or to expand in a traditional territorial sense through force. They are increasingly being brought into the mainstream of development by forces beyond their control, and into situations where traditional pastoral strategies may become less efficient. These strategies then become less useful to pursue".

Abolition of the local administrative systems without an effective alternative has subjected pastoral systems in the Sudan to extreme pressures and has caused drastic levels of deterioration of the rangelands (Abdullah 1982, Anon. 1982).

> "Transhumant nomads formerly fell within the native administrative system and were regulated through their tribal leadership, the *Nazir/Omda/Sheikh* structures, with jurisdiction over large territorial zones. The leaders understood the needs of the nomads and sedentaries and kept them apart. The new system (local government councils and a centralized administration) does not assign nomads to any local areas as they cross administration boundaries. The system does not lend itself to resolving the conflicts of these mobile people with their livestock. Land use under the former native administration was subject to tribal custom, but the new system now favours sedentaries through the law. For example, *gifar* (communal grazing reserves) are now open for any kind of investor to cultivate and no agency controls the use of former *gifar* land. Elimination of the native administration was so complete that it cancelled tribal boundaries, took away customary rules and broke down barriers between different tribal populations, causing rangeland havoc. It assisted selfish interests of individuals who took advantage of others in their group. There is an administrative gap here. . . . Since people are no longer obliged to follow customary law, conflicts between them became more acute. . . . There is no general policy to guide land use, no common law, and only unplanned expansion of agriculture goes on" (Abdullah 1982).

During the early part of the dry season (November–December), pastorals are fined when livestock trespass and damage unharvested crops (Teitelbaum 1984). Beside paying fines the nomads spend considerable time trying to locate the government corals. According to Teitelbaum (1984) even grazing standing crop residues late in the dry season is considered an offence and pastorals have to arrange deals with the sedentarized farmers or be subject to fines.

(b) *Socio-economic*

Traditionally, pastoral nomads raise large numbers of stock. Socially, livestock are signs of prestige and wealth. Pastorals are often accused of apparently irrational economic behaviour – principally refusal to dispose of stock which should be sold according to the usual economic standards (Pendelton and Van Dyne 1980). Under the traditional pastoral systems much of this behaviour could be attributed to rational and well calculated responses to the unpredictable marginal environments. Pastorals keep large numbers of livestock as a security against hazardous environment and natural disasters (UNESCO 1977). Nomads usually do not sell their livestock unless the price is right, and only when they need cash to buy food. Recently the prices of cattle which survived drought in the low-potential areas of Kenya were sold for as much as 10 times the price during the drought months (Mukhebi, pers. comm.). However, considering the marked increase in human population and the declining rangelands (in surface area and productivity) as a result of over-exploitation caused by the external factors discussed earlier, per capita livestock units have in fact declined below the subsistence requirements of many households.

(c) *Aridity*

Pastorals diversify the composition of their animals to cope with the environment of their grazing domains. For example, in the southern rangelands of Ethiopia, with relatively good reliable rainfall (600 mm), the cattle/small ruminant/camel ratio is 3.6/1.3/1, whereas in the much more arid northeast rangeland areas, the species ratio is 4.3/11.25/1. Both systems have species mix, but the reliance on small ruminants by the pastorals of the more arid regions indicates a greater drought risk (Cossins 1983). The significance of multi-species herding in marginal rangelands is illustrated here by comparing the range condition of Kitui area in eastern Kenya for cattle and goats (Skolvin 1980). Net rating situation arises from the fact that cattle, being grazers, seldom compete with goats which utilize considerable amounts of browse.

(d) *Desertification*

The definition offered by Rapp *et al.* (1976) states that

> "desertification, (desertization, desert encroachment) is the spread of desert-like conditions in arid and semi-arid lands. Desertification occurs as a result of: (1) severe and prolonged drought, and (2) man's over-exploitation of vegetation and soil in dry lands".

The man-made causes include overgrazing, over-cultivation, wood cutting and unprescribed fires. About 95% of the world's arid and semi-arid lands are threatened by desert encroachment (Mabbutt 1978). Large areas of pastoral grazing domains have been almost completely depleted during the last two decades. This has resulted in over-stocking and over-use of other marginal lands.

(e) *Deterioration of range productivity*

The following factors have led to the deterioration of the quantity and quality of range productivity.

1. *Overgrazing:* The arid and semi-arid lands hold the majority of the livestock in Africa and the Near East. Extremely high numbers of livestock are kept in rangelands of inherently low plant production. According to Le Houérou (1977), heavy continuous grazing at an average of twice the carrying capacity (2 ha/sheep) subjected the arid grazing lands of the Mediterranean region of Africa to an alarming state of degradation.
2. *Extended drought:* The recent Sahelian drought has inflicted heavy mortality upon pastoral livestock. Well-documented losses of 70% of herds have occurred in virtually all the arid and semi-arid regions of the developing world including Mongolia, Iran, Algeria, Soviet Kazakhstan, Botswana and the Sahelian countries (Sandford 1983).
3. *Cultivation of grazing lands:* The best conversion of primary elements into food for human consumption depends on the environment. It is generally accepted that the minimum annual rainfall necessary for permanent cropping is 500 mm. However, vast areas in the 200 and 300 mm precipitation zones were cultivated but with poor results (UNESCO 1978). For example, cereal areas in the Sahelian and Sudanian regions of Africa increased by 6% between 1960 and 1970 (UNESCO 1977). In North Africa and the Near East the increase was 10% in 20 years (1950–1970). Unfortunately, in these agriculturally marginal lands, the best pastoral rangelands (in feed quality) were the most targeted for food production. This practice has depleted the rangelands of valuable perennial forage species (Le Houérou 1977).
4. *Fires:* Rational (prescribed) fires are tools for range improvement. Traditionally, pastorals use fires to exterminate ticks and undesirable insects in the high-rainfall savanna zones (Majid 1972). Also, prescribed fires are useful for fire breaks and to revegetate new shoots during the early dry seasons. However, irrational fires during the dry season have the devastating effect of eliminating the forage biomass and seed reserves. Also, repeated burning replaces the high-quality perennials by the low quality annuals (Sidahmed and Koong 1984).
5. *Water points:* Water points have an adverse effect on marginal rangelands in two ways. For example, under-use in a marginal grazing area may result from the scarcity of drinking water available for livestock. On the other hand, bore holes located in dry-season grazing areas may attract more animals than the holding capacity of the range around the immediate perimeter of the water point, thus causing a permanent depletion of the seed bank of the desirable forage species. Pratt and Gwynne (1977) warned that

   > "great care is needed to supply water in over-abundance. Often there is a natural balance between livestock, water and grazing, arising from the fact that much of the water is seasonal, so that, as soon as one supply is exhausted, the stock are forced to move away to a fresh source before overgrazing occurs. Indiscriminate water development can very easily upset the natural balance".

6. *Abolition of local administration:* As discussed earlier, this factor usually favours the sedentarized agriculturists, cancels tribal boundaries and territorial grazing domains and eventually deals a final blow to the pastoral production system. Pastoralism and nomadism are traditional ways of life. It will be left to the politicians to prove and to the future to determine whether an alternative use of marginal lands is to be advocated. Until that is confirmed, ecologists will always look sympathetically to the traditional nomadic way of co-existence with such a fragile and extremely insecure environment.
7. *Poor animal performance:* Shortage and seasonal fluctuation of feed supply are the main constraints to animal production in the marginal ecosystems. The majority of the animals are poorly nourished and suffer from low resistance to disease and parasites. Also the cyclic pattern of migration subjects the stock to transmissible diseases. Because of tsetse fly-borne trypansomiasis, and mud during the rainy seasons, vast areas in Africa (between Lat. 10°N and 15°S) are not suitable for livestock production. Generally, productivity parameters are below the genetic potential of most of the pastoral livestock species.

RANGE IMPROVEMENT

Animal production in the extensive rangelands is constrained by natural (land, moisture and climate), biological (nutrition, health, parasites and predators), and socio-economic factors. Therefore, efforts to improve the complex grazing systems must consider all these production components. According to FAO (1974): "The grazing lands of the developing world are the most *over*-utilized and at the same time *under*-utilized resources in agriculture". The success of improvement decisions such as allocating land to stock owners or enforcing animal quotas is doubtful under the current political, economic and social situation of most of the developing countries.

The following list of operations and decisions to improve the grazing systems was adopted from Morley (1981) and Sidahmed and Koong (1984):

Range: Reseeding of desirable plants
Subdivision and resting
Feed consumption
Supplementary feeding
Water development
Fodder cultivation
Fire management

Animal: Herd composition
Stocking rate
Disease control
Offtake and marketing

The communal extensive grazing system prohibits fertilizer application. Also, range reclamation by fencing and rotational grazing is not financially feasible. Reseeding of viable desired species may be a possible strategy if strict grazing practices could be implemented.

The detailed discussion of the above points is beyond the scope of this paper. However, the following specific measures to improve feed supply are suggested:

1. Restriction of cultivation in the marginal lands to fodder cropping.
2. Intercropping feed with food crops in the semi-nomadic and transhumant farms. Generally, nomads do not farm, and the danger of over-exploitation is usually limited by scarcity of water in the arid regions.
3. Encouraging nitrogen-fixing, local leguminous trees and shrubs. This option is mostly uncertain to succeed in locations where phosphorous deficiency and competition with grasses prevail.
4. Wood cutting in the desert fringes (for fire or building of homesteads) should be strictly prohibited.

ANIMAL PRODUCTION SYSTEMS IN UAE – A CASE STUDY

(a) *Background*

The United Arab Emirates (UAE) is located between Lat. 23° and 26° N and Long. 52° and 56° E. The agro-ecological zone is typically very arid (Table 1). The majority of the landscape (92%) is flat plains and the country is surrounded to the east by the sea (the Gulf of Oman to the southeast and the Arabian Gulf to the northeast). The country has a hot and extreme climate with mean daily temperature ranging from 20°C in winter (December to February) to 30°C in summer. The hottest summer months are July and August (with maximum temperature of 45°C). Humidity is high and varies from 54% in October–April to 79% in May–September. Evaporation is generally very high, ranging between 2980 mm inland and 4050 mm at the coast.

(b) *Production systems*

Historically, the majority of UAE inhabitants were involved in livestock production in one way or another. While the nomads in the inner land raised camels and goats for milk, meat and cover (hair and hide), the coastal inhabiters traded in livestock imports from other countries. Crop residues and fresh fodder were used by the sedentary cultivators as feed for the small flocks raised in their household premises (UAE Ministry of Agriculture and Fisheries 1981). Basically the same system exists until now, although a marked increase in imports of live, fresh and processed meat has occurred during the last two decades.

(c) *Livestock species and breeds*

1. *Cattle:* Most of the cattle are from the Indian Zebu origin. Breeds introduced from Asia (Iran, Pakistan and the Sultanate of Oman) have more advantages in milk and meat production over the local breed which is generally small in size (average body weight of 250 kg) and produces 2 to 3 litres of milk per day. Although the exotic Freisians produce 3000 kg

of milk per year, they are limited to highly specialized dairy farms. Generally, Freisians and other European breeds are difficult to adapt to the hot environment of this region.

2. *Goats:* Goats are most suitable to the UAE desert environment, and are reared mainly for milk (Table 2). They include Indian, Iranian, Pakistani and Middle Eastern breeds. Local breeds are small in size and are characterized by black hair cover.

Table 2. Herd structures. UAE in 1980.*

Item	*Cattle*	*Sheep*	*Goats*	*Camel*
Total (herd)	25,665	132,237	341,622	58,709
Males (%)	38	20	15	15
Females (%)	62	80	85	84
Lactating (%)	43	52	57	71

*Source: UAE Ministry of Agriculture and Fisheries (1980).

3. *Sheep:* Sheep are raised for milk and meat. The majority of sheep flocks are raised on crop residues and by-products. Breeds include Awasi (from Syria) and Balushi (from Pakistan) which are good meat and wool producers. Local sheep are from the fat-tail type and are low milk producers. Somali sheep are imported into the country mainly for slaughter.
4. *Camels:* Camel population declined from 100,000 in 1969 to 45,000 in 1978 before increasing again in the subsequent years. Herd size per family ranges from 50–150 heads in the western parts of the country to 15–25 in the northern rangelands. Access to dry-season grazing was facilitated by the ability of the nomads to possess trucks and pick-ups. The procedure being followed is in such a way that the animals graze for several weeks on desert ranges (annual grasses, shrubs and scrub), while drinking water is truck-loaded once every two days. The Bedouins consider the camel as an important asset and a source of milk and meat. On average, about 4–5 litres of milk per head are produced, which is a reasonable amount under such very arid conditions. The husbandry of the lactating camel differs from that of the rest of the herd. Lactating camels graze during the day in the nearby pastures and are fed on concentrates during the evenings. The Arabian breed is the most dominant, with an average body weight of 500 to 700 kg and an annual milk production of 742 kg. Camels are also valued for recreation and as a traditional symbol of heritage.

(d) *Land Use*

Rangelands comprise only 2.6% (200,000 ha) of the total area. Out of 50,000–60,000 ha of potentially arable land only 23,000 are under cultivation (Table 3). This area is very small in spite of a threefold increase in the cultivated area (from 8,000 to 23,000 ha) between 1961–1965 and 1980 (FAO 1974, UAE Ministry of Agriculture and Fisheries 1981). The area

Table 3. Land use in the UAE.*

Item	*1961–1965*	*1967*	*1972*	*1977* (1000 ha)	*1978*	*1980*
Potentially arable	50–60	50–60	50–60	50–60	50–60	50–60
Cultivated	8	10	12	11	21	23
Rangeland	200	200	200	200	200	200
Forest and woodland	—	—	—	—	N.A.	25
Other	8152	8150	8147	8137	8111	
Total	8360	8360	8360	8360	8360	8360

*Source: FAO (1974), UAE Ministry of Agriculture and Fisheries (1981).
N.A. = Not available.

Table 4. Animal/human ratio in UAE.*

Ratio	*Year* 1970–1971	1980–1981
Cattle	0.07	0.02
Sheep	0.41	0.13
Goat	1.13	0.33
Camel	0.44	0.06

*Source: UAE Ministry of Agriculture and Fisheries (1981).

covered with forests and woody vegetation increased from 1,000 ha in 1977 to 25,000 in 1980. However, this increase in arable land was not large enough to provide self sufficiency in food production. Although the population density has increased significantly from 2.9 person/km^2 in 1970 to 13.5 person/km^2 in 1981, the animal/human ratio (Table 4) and area cultivated/human ratio declined. This decline is a reflection of the sharp increase in human population thus reducing the magnitude of the very significant increase achieved in animal numbers (Tables 5 and 6). The implications of this situation, human population boom and high dependency on external food production require considerable attention. This country has to be commended for a remarkable and genuine concern to protect its marginal lands and explore its water resources. The initially declining trend in livestock numbers (particularly camels) between 1969 and 1978 was

Table 5. Livestock numbers, UAE.*

Year	*Cattle*	*Sheep*	*Goats*	*Camel*
1969–1971	16,000	94,000	257,000	100,000
1976	25,000	103,000	270,000	60,000
1977	19,000	95,000	250,000	48,000
1978	15,000	90,000	240,000	45,000
1979	23,332	120,216	310,566	53,372
1980	25,665	132,237	341,622	58,709

*Source: FAO 1974, UAE Ministry of Agriculture and Fisheries (1980).

Table 6. Number of animals, and farmers receiving production incentives in UAE, AL-Ain Agriculture and Animal Production Department.*

Year	*Small ruminants (head)*	*Camels (head)*	*Farmers*	*Incentives, UAE Dirham† $\times 10^6$*
1979	102,149	30.680	2,902	11.24
1981	109,549	36,244	3,222	12.73
1983	156,646	44,062	3,551	16.64
1984	187,655	50,550	N.A.	N.A.

*Source: UAE Emirate of Abu Dhabi, Agriculture and Animal Production Department (1983)

†US$1 = 3.6 UAE Dirham, January 1985.

reversed during the recent few years (Tables 5 and 6). For example, the Department of Agriculture and Animal Production at Al-Ain provides farmers with incentives to raise livestock ($55/yr for each camel and $14/yr for each sheep or goat). This was reflected in an increase of 15% in the population of camels (from 44,062 to 50,550) and of 20% in small ruminant population (from 156,646 to 187,655 heads) between 1983 and 1984 (El-Darmaky, pers. comm.). Also these incentives have been reflected in the increase of participating farmers from 2,902 in 1979 to 3,551 in 1983. However, the marked increase in the amounts of money paid (increase of 48% between 1979 and 1983) compared to the number of recipients, which showed a 22% increase, should be of major concern particularly as the country's oil revenues are currently declining. The agricultural labour force of the UAE (12,000 people) is very small compared to the total human population of 1,040,000. In the high-potential temperate regions, small proportions of the total population are engaged in food production and the production usually exceeds local consumption. It would be misleading to draw comparisons between the temperate high-potential lands and the marginal ecosystems. Even when the potentially arable lands in this very arid region are fully exploited, the output will not be sufficient for local consumption. A complementary formula of collaboration between regions of diverse resources, e.g. a unified confederation between states and mutual trade agreements is essential to secure an even distribution of resources.

LITERATURE CITED

Abdullah, Suliman H. 1982. Problems of nomads in the Nuba Mountains. Memo. Min. Agric. Nat. Resources, Kordofan Region, Sudan.

Anon. 1982. Environmental context of development in the Sudan. Geogr. Dept., Univ. of Khartoum, with assistance from the program for Int. Devt., Clark Univ., Worchester, Ma., USA.

Cossins, Noel. 1983. Production strategies and pastoral man, p. 213–231. *In:* Proc. Workshop on Pastoral System Research in Sub-Saharan Africa. ILCA (Int. Livestock Centre for Africa), 21–24 March, Addis Ababa, Ethiopia.

Dyson-Hudson, N. 1982. Changing production strategies in East African range livestock systems: An overview. Informal paper, ILCA (Int. Livestock Centre for Africa), Addis Ababa, Ethiopia.

FAO. 1974. FAO production yearbook, vol. 32, Rome.

French, M. H. 1966. Animal production in Savanna areas. FAO, LA: SF/SD/66/8, Rome.

Hallam, David. 1983. Livestock development planning: A quantitative framework. Centre for Agricultural Strategy (CAS) No. 12, Univ. of Reading, UK.

Hjort, A. 1976. Traditional land use in marginal dry lands, p. 24–43. *In:* A. Rapp, H. N. Le Houerou and B. Lundholm (eds), Can desert encroachment be stopped? A study with emphasis on Africa. Ecol. Bull. No. 24, Swed. Nat. Sci. Res. Coun., NFR, Stockholm.

Johnson, D. L. 1979. Management strategies for dry lands: Available options and answered questions, p. 26–35. *In:* J. A. Mabbutt (ed), Proc. Khartoum Workshop on Arid Lands Management. UN University Press, Tokyo.

King, J. 1983. Livestock water needs in pastoral Africa in relation to climate and forage. ILCA (Int. Livestock Centre for Africa), Res. Rep. No. 7, Addis Ababa, Ethiopia.

Le Houerou, H. N. 1977. The rangelands of the Sahel. *J. Range Manage.* **33** (1): 41–46.

Lundholm, B. 1976. Domestic animals in arid lands, p. 29–42. *In:* A. Rapp, H. N. Le Houerou and B. Lundholm (ed), Can desert encroachment be stopped? A study with emphasis on Africa, Ecol. Bull. No. 24, Swed. Nat. Sci. Res. Coun., NFR, Stockholm.

Mabbutt, J. A. 1978. Desertification of Australia in its global context. *Search* **9:** 252–256.

Majid, R. A. 1972. Towards an efficient range and pasture management in southern Sudan. Min. Agric., Southern Region, Juba.

Morley, F. H. W. 1981. Management of grazing systems, p. 379–400. *In:* F. H. W. Morley (ed), *World animal science, B1: Grazing animals.* Elsevier, Amsterdam.

Pendelton, D. F. and G. M. Van Dyne. 1980. Livestock and world grazing lands – status and outlook. A working paper developed in US Nat. Sci. Found. Proj. Prediction of grazing land productivity under climate variations. Colorado State Univ., Fort Collins, USA.

Pratt, D. J. and M. D. Gwynne (ed). 1977. *Rangeland management and ecology in East Africa.* Hodder and Stoughton, London.

Rapp, A., H. N. Le Houerou and B. Lundholm (ed). 1976. Can desert encroachment be stopped? A study with emphasis on Africa. Ecol. Bull. No. 24, Swed. Nat. Sci. Res. Coun. NFR, Stockholm.

Sandford, Stephen. 1983. Organization and management of water supplied in Tropical Africa. ILCA (Int. Livestock Centre for Africa), Res. No. 8, Addis Ababa, Ethiopia.

Sidahmed, A. E. and J. L. Koong. 1984. Application of systems analysis to nomadic livestock production in the Sudan, p. 61–76. *In:* J. R. Simpson and P. Evangelor (ed), *Livestock development in sub-Saharan Africa: constraints, prospects and policy.* Westview Press, Boulder, Colorado, USA.

Sidahmed, A. E. and J. G . Morris. 1982. Foraging strategies of Spanish goat browsing California chaparral. Proc. III Int. Goat Symp., Tucson, Arizona.

Skolvin, J. M. 1980. Some aspects of range research and development in the semi-arid zones of Kenya. UNDP-FAO, AGP: SF/KEN/11, Tech. Rep. No. 1.

Society for Range Management. 1974. *A glossary of terms used in range management.* SRM, Denver, Colorado, USA.

Stiles, Daniel N. 1983. Description and pastoral development in northern Kenya. *Nomadic People* **13:** 1–14.

Teitelbaum, J. M. 1984. The transhumant production system and change among Hawazma nomads of the Kordofan Region, Western Sudan. *Nomadic People* **15:** 51–65.

UNESCO. 1977. Development of arid and semi-arid lands: Obstacles and prospects, MAB Tech. Notes 6, UNESCO, Paris.

UNESCO. 1978. MAB Mediterranean Scientific Conference: Regional Meeting for MAB National Committees of Countries Bordering the Mediterranean Sea. Final Report of Conference, 26 September–2 October 1976. Montpellier, UNESCO, MAB Report Series No. 43.

United Arab Emirates, Emirate of Abu Dhabi Department of Agriculture and Animal Production. 1983. Ann. Statist. Bull. (in Arabic), Al-Ain, UAE.

United Arab Emirates Ministry of Agriculture and Fisheries. 1980. Ann. Statist. Bull. (in Arabic), Kazim Press, Dubai.

United Arab Emirates Ministry of Agriculture and Fisheries. 1981. Agriculture (in Arabic). El-Matbaa' El-Fanniya, Dubai, UAE.

Winrock International. 1983. Sheep and goats in developing countries: their present and potential role. A World Bank Tech. Paper Presented by Winrock International, The World Bank, Washington, D.C., USA.

٢ – التعرف على أنواع النباتات الملائمة وفقا لظروف المنطقة.
٣ – إدخال أنواع الأعلاف التي تتحمل الجفاف والملوحة.
٤ – تطوير زراعة النباتات العلفية والمراعي من أجل تأمين الإحتياطي العلفي.
٥ – إجراء دراسات المراعي المروية.
٦ – القيام بدراسات الحشائش والإستجابة للرعي والاجهاد.
٧ – الإستمرار في دراسات الحشائش والشجيرات المحلية المعمرة.
٨ – التعرف على المواد الوراثية للبذور والمحافظة على الأنواع الجيدة.
٩ – توحيد أساليب وطرق قياس ومراقبة النباتات الرعوية.
١٠ – تجميع المعلومات المدونة عن النباتات المحلية في الخليج العربي.

ب – الأبحاث الخاصة بالإنتاج الحيواني:

١ – دراسة السلوك الرعوي ليشمل استساغة النباتات من قِبل الحيوان، وتحركات قطعان الماشية، والأنظمة الغذائية.
٢ – دراسة تركيب وديناميكية الحيوان الرعوي لضمان إنتاجية عالية.
٣ – التعرف على طبيعة الأجناس المحلية والمستجلبة من كافة أنواع الحيوانات.
٤ – التأكيد على تغذية وإنتاج الجمال والماعز والأغنام.
٥ – القيام بدراسات التغذية لموازنة التموين العلفي طوال العام.
٦ – دراسة ديناميكية الثروة الحيوانية والوضع الراهن للحيوانات البرية.

ج – الابحاث الخاصة بنظم المراعي:

١ – استعمال نظم الرعي المتكاملة من أجل تقييم النباتات والحيوانات.
٢ – دراسة الرعي والعوامل الإيكولوجية لضمان الحد الاقصى من الإنتاج المستمر.
٣ – تطوير نظم الرعي الملائمة لتقاليد المجتمع وإيكولوجية المراعي.
٤ – إجراء دراسات الحمولة الرعوية.
٥ – إجراء الدراسات الاقتصادية لتقييم نظم الرعي المتكامل.

د – الدراسات الإجتماعية والإقتصادية:

إن الأراضي الرعوية في المناطق القاحلة من العالم تعتبر مصدرا للرزق للمجتمعات الرعوية وفقا لإنتاجيتها البيولوجية، ويجب المحافظة على نوعية الحياة وذلك بضمان عدم حدوث مزيد من التدهور وإيقاف عمليات التصحر مع التأكيد على تشجير المناطق المتصحرة، وبتلبية الحاجات الإجتماعية للمجتمعات. كذلك ينبغي المحافظة على التراث الوطني والإقليمي لهذه المجتمعات، ولا يتأتى ذلك إلا بتفهم أفضل لحاجات مجتمعات البدو والحاجات الترفيهية لسكان المدن. وعليه يوصى بالآتي:
١ – دعم النظم التقليدية للرعي مع التركيز على تنقلات المجتمعات البدوية.
٢ – استحداث أساليب مقبولة لتحسين المراعي والممارسات الرعوية.
٣ – دراسة مفهوم ملكية الأراضي.

٤ – إجراء الأبحاث حول السلوك الترفيهي مثل :
* قيادة السيارات خارج الطرق المعبدة .
* الصيد غير المنظم .
* إقتلاع النباتات .

توصية رقم (٢) : سبل تطوير وإدارة المراعي :

أ – على كل قطر تطوير نظم مقبولة اجتماعيا وذلك لتنظيم عمليات الرعي .
ب – يجب وضع نظام إقليمي لمراقبة الرعي .
ج – على كل قطر أن يضع الأولويات الخاصة بالإستفادة من الأراضي الرعوية .
د – يجب إنشاء مستودعات لحفظ المياه وذلك للإستفادة من الأراضي الرعوية الجافة بصورة موسمية وكذلك إقامة مستودعات لتأمين الاحتياطي العلفي .
هـ – تحسين وتطوير المراعي المتدهورة .
و – استعمال التقنيات المناسبة لتحسين المراعي .

توصية رقم (٣) : تشجيع الإلتحاق بالمعاهد والجامعات لدراسة المراعي :

أ – إدخال علم إدارة المراعي ضمن مناهج بعض الجامعات في منطقة الخليج العربي .
ب – تدريس علوم إدارة المراعي في المعاهد الفنية .
ج – إختيار بعض المواقع المناسبة واستعمالها كمراكز تعليم للممارسات الملائمة لتطوير المراعي .
د – تنظيم دورات دراسية قصيرة على نطاق إقليمي ، ويكون ذلك بصورة مبدئية في محطة أبحاث سكاكا بالمملكة العربية السعودية وبمعهد الكويت للأبحاث العلمية بالتنسيق مع الأمانة العامة لمجلس التعاون لدول الخليج العربية .
هـ – إرسال الطلاب من المنطقة في بعثات للدراسات العليا ، ويجب تطوير برنامج دراسي خاص للمراعي بمجلس التعاون لدول الخليج العربية .
و – إنشاء مراكز تدريبية ومدارس للتعليم الزراعي بمناطق القبائل البدوية .
ز – وضع برامج للتوعية على المستويين الوطني والإقليمي مع وضع حوافز من الدولة لدراسة علم إدارة المراعي .

توصية رقم (٤) : التعاون الإقليمي والعالمي في ادارة المراعي :

أ – إتخاذ إجراءات عاجلة ووضع خطة للمتابعة من قبل كل الأقطار المعنية وتكوين لجنة لمتابعة تنفيذ التوصيات .
ب – عقد المؤتمر العالمي الثاني حول إدارة المراعي في الخليج العربي خلال سنتين .
ج – على كل قطر تحديد جهة خاصة تكون مسؤولة عن إدارة المراعي .
د – وضع دليل للمؤسسات والأشخاص ومشاريع البحث الخاصة بادارة المراعي .
هـ – تكوين جمعية لإدارة المراعي أو تكوين منظمة إقليمية لهذا الغرض .
و – البدء في عملية تبادل البذور المحسنة .
ي – تطوير عملية تبادل الخبرات والمعلومات في مجال إدارة المراعي .

التوصيات والقرارات

الكويت ٢٢ – ٢٤ ابريل ١٩٨٥

يعبر المشاركون في المؤتمر العالمي الأول حول إدارة المراعي في الخليج العربي المنعقد في الكويت في الفترة من ٢٢ – ٢٤ ابريل ١٩٨٥ في فندق حياة ريجنسي عن أسمى آيات الشكر والتقدير الى حكومة الكويت لاستضافتها هذا المؤتمر الهام، كما يوجهون الشكر الى الهيئات المنظمة للمؤتمر وهي: معهد الكويت للأبحاث العلمية ومؤسسة الكويت للتقدم العلمي ومجلس حماية البيئة، وذلك لدعمها وتنظيمها لهذا المؤتمر. ويتوجهون أيضا بالشكر والتقدير لمعهد الكويت للأبحاث العلمية لمبادرته بطرح الفكرة والتنسيق مع الهيئات الاخرى في الإعداد والتحضير لهذا المؤتمر. ويوصى المؤتمرون بما يأتي:

التوصيات العامة

توصية رقم (١):

يقر المؤتمر العالمي الأولى حول إدارة المراعي بأهمية الأراضي الرعوية في منطقة الخليج العربي والتي تشكل أكثر من ٩٠٪ من مساحة المنطقة، إذ أن هذه الموارد تمثل موارد متجددة كالحيوان والنبات والألياف، كما تمثل منتجعات للراحة والإستجمام. ويقر المؤتمر أيضا بأن حالة المراعي في المنطقة تتدهور بسرعة كبيرة نتيجة لسوء استغلال الموارد الطبيعية مما سيؤدي إلى نتائج خطيرة مثل تعرية التربة والتصحر ونضوب موارد النبات والتربة. لذا يوصي المؤتمر ببذل الجهود المخلصة من أجل الوصول الى الإنتاجية الحقيقية الممكنة للمراعي. كما يؤكد المؤتمر بأن الجهود الحالية في مجال إدارة المراعي بحاجة إلى مزيد من التكثيف والدعم والتشجيع.

توصية رقم (٢):

نظرا لأهمية موارد المياه في المناطق القاحلة وشبه القاحلة وخاصة الامطار، كونها العامل الأساسي في تحديد الإنتاجية، لذا يوصي المؤتمر بالدراسة الشاملة والتفهم الكامل لهذا المورد الهام من أجل وضع سياسات الادارة وخطط استعمال المياه واستغلالها على أفضل وجه من الكفاءة حتى يمكن الحفاظ عليها.

التوصيات الخاصة

توصية رقم (١): الابحاث:

أ – الأبحاث الخاصة بالإنتاج النباتي:

١ – القيام بحصر ومراقبة الإنتاجية الفعلية والمحتملة للنباتات وإجراء مسح شامل لها.

ويسعدنا أن نعبر عن شكرنا وتقديرنا لرؤساء جلسات المؤتمر: السيد/نزار حسين، والدكتور/محمد العطار، والدكتور/ثاديس بوكس، والدكتور/عادل سلمان، والمهندس/فهد الحساوي، والدكتور/ ج. جودين، والدكتور/محمد عبد الرزاق. كما نوجه شكرنا الخاص لمقدمي البحوث الذين أسهموا في إنجاح المؤتمر، وإلى أعضاء لجان المؤتمر: اللجنة التنظيمية، واللجنة العلمية، ولجنة صياغة التوصيات والقرارات، الذين كان لجهودهم القيمة خير الأثر في إدارة المؤتمر بكفاءة ونجاح.

سميرة عمر　　　　فيصل طه　　　　رياض الحلوجي

مقدمة

تحت رعاية سمو ولي العهد رئيس مجلس الوزراء الشيخ سعد العبد الله السالم الصباح ، أقيم في فندق حياة ريجنسي بالكويت المؤتمر العالمي الأول حول إدارة المراعي في الخليج العربي خلال الفترة من ٢٢ – ٢٤ أبريل ١٩٨٥ . وقد قام بتمويل المؤتمر وتنظيمه ثلاث جهات هي : معهد الكويت للأبحاث العلمية ، ومؤسسة الكويت للتقدم العلمي ، ومجلس حماية البيئة .

نظرا لأهمية تطوير المراعي وتحسينها في منطقة الخليج فقد عقد المؤتمر من أجل تحقيق الأهداف التالية : تبادل الأفكار والخبرات بين العلماء والمختصين العرب والأجانب في مجال تطوير إدارة المراعي في الخليج العربي والمناطق الجافة ووضع توصيات أولية في مجال إدارة الأراضي الجافة ، والبحث في إختيار التقنيات والوسائل المتبعة في إدارة المراعي وتطبيقها بما يتلائم مع الظروف المحلية ، وتشجيع البحث العلمي الدقيق مع وضع استراتيجية متكاملة لتحديد الأولويات والسبل للتغلب على المشاكل والمعوقات التي تعترض المراعي ، وأخيرا إيجاد علاقات علمية متطورة بين مؤسسات البحث العلمي والهيئات التي تعني بتحسين المناطق الرعوية في منطقة الخليج ، وحثها على تبادل البحوث والخبرات فيما بينها .

وأناب سمو ولي العهد رئيس مجلس الوزراء الشيخ سعد العبد الله السالم الصباح سعادة وزير الدولة لشؤون مجلس الوزراء الأستاذ راشد عبد العزيز الراشد في افتتاح المؤتمر حيث ألقى كلمة الرعاية . وبعد ذلك ألقى رئيس مجلس إدارة الهيئة العامة لشؤون الزراعة والثروة السمكية الشيخ ابراهيم الدعيج الصباح كلمة بين فيها أهمية المراعي وإمكانية تطويرها . ثم ألقى بعد ذلك عميد كلية الموارد الطبيعية في جامعة يوتا الأمريكية الدكتور/ثاديس بوكس المقدمة العلمية للمؤتمر .

اشتمل برنامج المؤتمر على خمس جلسات علمية وجلسة خاصة لتقديم الأبحاث في صورة ملصقات . كذلك نظمت إدارة المؤتمر رحلة حقلية إلى الشركة المتحدة للإنتاج الزراعي ومحطة أبحاث المراعي بالصليبية التابعة لمعهد الكويت للأبحاث العلمية . وتم إلقاء ٢١ بحثا في الجلسات العلمية قدمت من دول مجلس التعاون الخليجي والدول العربية الشقيقة سوريا والعراق ومصر الى جانب الدول الأجنبية مثل المكسيك وأمريكا ، ومنظمة الأغذية والزراعة الدولية التابعة للأمم المتحدة ، والمركز الدولي للزراعة في المناطق الجافة . وقد اختير من بين البحوث ١٧ بحثا تم نشرها في هذا المجلد ، الذي تحملت نفقات طباعته مؤسسة الكويت للتقدم العلمي .

وفي حفل الختام تم مناقشة التوصيات والقرارات والتي تبرز أهمية المراعي الطبيعية في منطقة الخليج ، وأهمية الأبحاث الخاصة لإدارة المراعي وسبل تطويرها إلى جانب البحث في تطوير الدراسات والتعاون الأقليمي في مجال المراعي .

لجنة صياغة التوصيات والقرارات

حيوانات المراعي

المؤتمر العالمي الأول
حــول
إدارة المراعي في الخليج العربي

الكويت ٢٢ – ٢٤ ابريل ١٩٨٥

المحتويات

اللجنة العلمية

فيصل خضر طه	(معهد الكويت للأبحاث العلمية)	رئيساً
عمر أبو الشوارب	(الهيئة العامة لشؤون الزراعة والثروة السمكية)	مقرراً
سميرة السيد عمر	(معهد الكويت للأبحاث العلمية)	
محمد عبد الرزاق	(معهد الكويت للأبحاث العلمية)	
رياض الحلوجي	(جامعة الكويت)	
عادل سلمان	(معهد الكويت للأبحاث العلمية)	

اللجنة العلمية

لجنة صياغة التوصيات والقرارات

ثاديس بوكس	(الولايات المتحدة الأمريكية)	رئيساً
عبده الشريف	(المملكة العربية السعودية)	رئيسا مشاركا
أحمد النقيب	(الكويت)	
فهد الحساوي	(الكويت)	
فيصل طه	(الكويت)	
مصطفى الدسوقي	(الكويت)	
سميرة عمر	(الكويت)	
سالم الشنفري	(عمان)	
سعد الكواري	(قطر)	
عبد الرحمن الداوود	(الامارات العربية المتحدة)	
عبد المولى صالح	(الامارات العربية المتحدة)	
ج. د. جودين	(الولايات المتحدة الأمريكية)	

المؤتمر العالمي الأول
حـول
إدارة المراعي في الخليج العربي

الكويت ٢٢ – ٢٤ ابريل ١٩٨٥

الجهـات المنظمـة

معهد الكويت للأبحاث العلمية
مؤسسة الكويت للتقدم العلمي
مجلس حماية البيئة ، الكويت

اللجنة التنظيمية

سميرة السيد عمر	(معهد الكويت للأبحاث العلمية)	رئيساً
وفاء الخميس	(مجلس حماية البيئة)	مقررا
عدنان العبد المحسن	(مؤسسة الكويت للتقدم العلمي)	
فيصل خضر طه	(معهد الكويت للأبحاث العلمية)	
فوزية السديراوي	(معهد الكويت للأبحاث العلمية)	
ساهرة الدوسري	(معهد الكويت للأبحاث العلمية)	
عبد الله طاهر عمر	(معهد الكويت للأبحاث العلمية)	

طبع لأول مرة في سنة ١٩٨٩
كيجان بول العالمية المحدودة
لندن ، انجلترا

الموزعون :
چون وايلي وأولاده المحدودة
انجلترا

روتلدج وتشابمان وهول
نيويورك
الولايات المتحدة الامريكية

الرقم المسلسل العالمي للكتاب
٠٧١٠٣٠٣٦٠٢

الإتجاهات الحديثة في إدارة المراعي
في الأراضي الجافة

وقـائـع
المؤتمر العالمي الأول
حول إدارة المراعي في الخليج العربي

رياض الحلوجي
جامعة الكويت

فيصل خضر طه **سميرة السيد عمر**
معهـد الكـويت للأبحـاث العلمية

كيجان بول العالمية
بالتعاون مع
مؤسسة الكويت للتقدم العلمي

الإتجاهات الحديثة في إدارة المراعي
في الأراضي الجافة